# Aging Research - Methodological Issues

Carmen García-Peña
Luis Miguel Gutiérrez-Robledo
Mario Ulises Pérez-Zepeda
Editors

# Aging Research - Methodological Issues

Second Edition

Springer

*Editors*
Carmen García-Peña
National Institute of Geriatrics
Mexico City, Mexico

Luis Miguel Gutiérrez-Robledo
National Institute of Geriatrics
Mexico City, Mexico

Mario Ulises Pérez-Zepeda
National Institute of Geriatrics
Mexico City, Mexico

ISBN 978-3-030-07025-0 ISBN 978-3-319-95387-8 (eBook)
https://doi.org/10.1007/978-3-319-95387-8



This Springer imprint is published by the registered company Springer Nature Switzerland AG
The registered company address is: Gewerbestrasse 11, 6330 Cham, Switzerland

# Preface to the Second Edition

This second edition of *Aging Research - Methodological Issues* is presented after 3 years of our first appearance. Drivers of the first edition are the same for this second edition and remain as relevant as in the first one. Aging population process represents the most important demographic issue in the world and particularly in low and middle economies that are currently facing several challenges in the social, economic, welfare, and health services dimensions among others.

The second edition maintains the idea of doing a book with the aim of integrating crucial features in aging research, such as multimorbidity, frailty, function, cognition, healthy aging with the principles of research methodology.

With this in mind, this new edition retains the organization and the general structure of the previous one. The twelve previous chapters were reviewed and updated. The review of the scientific method is presented again as the first chapter, after the introduction, and it includes a new discussion about complex systems applied to human aging. Biomedical research in aging is now written by a team of expert researchers that introduce new lines of investigation and future perspectives. As in the previous edition, classical research designs were included in the first chapters, including descriptive studies, case-control studies, longitudinal studies and clinical trials as well as systematic reviews, with updated information. Qualitative research and mixed methods are now presented by international authors with enormous experience in these topics. Chapter 15 is a discussion about the transference of health research results into aging policy; focusing in the urgent need of evidence in all the health systems to make better decisions in the aging field, taking advantage of what research provides to stakeholders. The discussion about the relationship between technology and aging was also included, with special emphasis on ubiquitous sensing, a continuously growing field both in engineering and aging.

Six new chapters were included: Chapter 4 dedicated to Geroscience which is a modern and emerging discipline based on finding connections between the "hallmarks of aging." Chapter 12 is focused on health systems research in aging. Health services have been particularly challenged due to an increase of health demands but also of a lack of scientific evidence. We are confident that this chapter will improve the understanding of how societies have to respond to the aging process. Big data and data mining are discussed in Chap. 14. Both are powerful tools to obtain information that could be used to improve the health status of older people. Ethical considerations in aging research are presented in Chap. 16. This chapter argues that such exceeding medical research should always be accompanied by an ethical stance, specifically focusing on aging population. The ethical stance in research serves to, first and foremost, look to safeguard the dignity of those it researches. Chapter 17 presents a crucial topic, the process involved with searching for aging research funding. Very specific key points are presented in order to write and present a successful grant proposal when focusing on the aging field. Finally, Chap. 18 is focused on discussion of the future of aging research, and how we need to move from disease paradigms to understand the person with a holistic perspective.

We have to say that many aspects of this book have not changed. It was written with several audiences in mind. We hope that under- and post-graduate students who are interested in aging research for the first time find this book challenging and useful. Senior researchers that have not done research in the area also can find a different perspective, and refreshing concepts may be found all over the diverse chapters.

Aging research must be as a top priority of any national research agenda. As in other medical branches, researchers need to be well trained and prepared; enough funds and institutional supports are needed to obtain sounding data that has the potential to impact how older adults are taken care of in all areas of the society.

After all, obtaining results with a standardized methodology will lead in turn to the formulation of new questions that will continue enriching the ever-growing field of aging research. We hope that this book will aid in achieving these goals.

National Institute of Geriatrics
Mexico City, Mexico

Carmen García-Peña
Luis Miguel Gutiérrez-Robledo
Mario Ulises Pérez-Zepeda

# Contents


1 **The Need for Differentiated Research Methodology in Aging** ...... 1
Mario Ulises Pérez-Zepeda, Carmen García-Peña, and Luis Miguel Gutiérrez-Robledo

2 **The Scientific Method as a Point of Departure in Aging Research** ...... 11
Ruben Fossion and Leonardo Zapata-Fonseca

3 **Biomedical Research in Aging** ...... 25
José Mario González-Meljem, Scott Haston, Suchira Gallage, and Andrew J. Innes

4 **Geroscience** ...... 55
Isabel Arrieta-Cruz and Armando Luna-López

5 **Descriptive Studies in Clinical Gerontology and Geriatrics** ...... 63
Mario Ulises Pérez-Zepeda, Lorena Jocabed Rocha Balcázar, and Miguel Germán Borda

6 **Qualitative Research in Gerontology and Geriatrics** ...... 73
Fernando A. Wagner, Laurens G. Van Sluytman, Halaevalu F. Ofahengaue Vakalahi, and Chioma Nwakanma Wosu

7 **Case-Control Studies in Aging Research** ...... 83
Sergio Sánchez-García, Erika Heredia-Ponce, Luis Pablo Cruz-Hervert, Ángel Cárdenas-Bahena, Laura Bárbara Velázquez-Olmedo, and Carmen García-Peña

8 **Longitudinal Studies and Older Adults Cohorts** ...... 95
Carmen García-Peña, Claudia Espinel-Bermúdez, Pamela Tella-Vega, Mario Ulises Pérez-Zepeda, and Luis Miguel Gutiérrez-Robledo

**9 Clinical Trials on Aging Research** . . . . . . . . . . . . . . . . . . . . . . . . . . . . . 115
Mario Ulises Pérez-Zepeda, Antonio Cherubini, Carmen García-Peña, Elisa Zengarini, and Luis Miguel Gutiérrez-Robledo

**10 Mixed Methods in Geriatrics and Gerontology Research** . . . . . . . . . . 129
Joseph J. Gallo and Jin Hui Joo

**11 Systematic Reviews and Meta-Analysis in Aging Research** . . . . . . . . 143
Miguel Ángel Villasís-Keever, Mario Enrique Rendón-Macías, and Raúl Hernán Medina-Campos

**12 Health Systems Research in Aging** . . . . . . . . . . . . . . . . . . . . . . . . . . . 157
Hortensia Reyes-Morales, Svetlana V. Doubova, and Ricardo Pérez-Cuevas

**13 Technology and Aging: Ubiquitous Sensing Technology for Aging Research** . . . . . . . . . . . . . . . . . . . . . . . . . . . . . . . . . . . . . . . . . 175
Jesús Favela and Luis A. Castro

**14 The Challenge of Big Data and Data Mining in Aging Research** . . . . 185
Juan Carlos Gómez-Verján and Luis Miguel Gutiérrez-Robledo

**15 Research in Public Policies for Aging** . . . . . . . . . . . . . . . . . . . . . . . . 197
Elizabeth Caro-López and Ernesto Velasco-Sánchez

**16 Ethical Issues in Research in Aging** . . . . . . . . . . . . . . . . . . . . . . . . . 209
Tirso Zúñiga-Santamaría and Carmen Jimena Vázquez-García

**17 Integration of Consortiums and Search for International Funding** . . . . . . . . . . . . . . . . . . . . . . . . . . . . . . . . . . 221
David X. Marquez, Iraida V. Carrion, Susan Aguiñaga, and Melissa Lamar

**18 Future of Aging Research** . . . . . . . . . . . . . . . . . . . . . . . . . . . . . . . . . . 231
Matteo Cesari, Marco Canevelli, and Mario Ulises Pérez-Zepeda

**Index** . . . . . . . . . . . . . . . . . . . . . . . . . . . . . . . . . . . . . . . . . . . . . . . . . . . . . 243

# Contributors

**Susan Aguiñaga** Department of Kinesiology and Public Health, University of Illinois at Urbana-Champaign, Champaign, IL, USA

**Isabel Arrieta-Cruz** Department of Basic Research, National Institute of Geriatrics, Mexico City, Mexico

**Lorena Jocabed Rocha Balcázar** Internal Medicine Department, Local General Hospital Number 27, Mexican Institute of Social Security, Mexico City, Mexico

**Miguel Germán Borda** Aging Institute, Faculty of Medicine, Pontificia Universidad Javeriana, Bogotá, Colombia

Centre for Age-Related Diseases, Stavanger University Hospital, Stavanger, Norway

**Marco Canevelli** Department of Human Neuroscience, "Sapienza" University, Rome, Italy

**Ángel Cárdenas-Bahena** Research Unit in Epidemiology and Health Services, Aging Area, National Medical Center Century XXI, Mexican Institute of Social Security, Mexico City, Mexico

**Elizabeth Caro-López** National Institute of Geriatrics, Mexico City, Mexico

**Iraida V. Carrion** School of Social Work, University of South Florida, Tampa, FL, USA

**Luis A. Castro** Department of Computing and Design, Sonora Institute of Technology (ITSON), Ciudad Obregon, Sonora, Mexico

**Matteo Cesari** Geriatric Unit, Fondazione IRCCS Ca' Granda - Ospedale Maggiore Policlinico, University of Milan, Milan, Italy

**Antonio Cherubini** Istituto Nazionale di Ricovero e Cura per Anziani, Ancona, Italy

**Luis Pablo Cruz-Hervert** Center for Research on Infectious Diseases, National Institute of Public Health, Mexico City, Mexico

**Svetlana V. Doubova** Epidemiology and Health Services Research Unit "Centro Médico Nacional Siglo XXI", Mexican Institute of Social Security, Mexico City, Mexico

**Claudia Espinel-Bermúdez** Research Unit in Clinical Epidemiology, West Medical Center, Mexican Institute of Social Security, Guadalajara, Jalisco, Mexico

**Jesús Favela** Computer Science Department, Center for Scientific Research and Higher Education of Ensenada, Ensenada, Baja California, Mexico

**Rubén Fossion** Nuclear Sciences Institute and Centre for Complexity Science (C3), National Autonomous University of Mexico, Mexico City, Mexico

**Suchira Gallage** Division of Chronic Inflammation and Cancer, German Cancer Research Center (DKFZ), Heidelberg, Germany

**Joseph J. Gallo** Mixed Methods Research Training Program for the Health Sciences, Department of Mental Health, Johns Hopkins Bloomberg School of Public Health, Baltimore, MD, USA

**Carmen García-Peña** Research Director, National Institute of Geriatrics, Mexico City, Mexico

**Juan Carlos Gómez-Verján** Department of Basic Research, National Institute of Geriatrics, Mexico City, Mexico

**José Mario González-Meljem** Department of Basic Research, National Institute of Geriatrics, Mexico City, Mexico

**Luis Miguel Gutiérrez-Robledo** Director, National Institute of Geriatrics, Mexico City, Mexico

**Scott Haston** Developmental Biology and Cancer Research Programme, UCL Great Ormond Street Institute of Child Health, London, UK

**Erika Heredia-Ponce** Department of Public Health and Oral Epidemiology, Faculty of Dentistry, National Autonomous University of Mexico, Mexico City, Mexico

**Andrew J. Innes** MRC London Institute of Medical Sciences (LMS), London, UK

**Jin Hui Joo** Department of Psychiatry and Behavioral Sciences, Johns Hopkins University School of Medicine, Baltimore, MD, USA

**Melissa Lamar** Rush Alzheimer's Disease Center, Department of Neurological Sciences, Rush Medical College, Chicago, IL, USA

**Armando Luna-López** Department of Basic Research, National Institute of Geriatrics, Mexico City, Mexico

**David X. Marquez** Department of Kinesiology and Nutrition, University of Illinois at Chicago, Chicago, IL, USA

Rush Alzheimer's Disease Center, Chicago, IL, USA

**Raúl Hernán Medina-Campos** Department of Geriatric Epidemiology, National Institute of Geriatrics, Mexico City, Mexico

**Halaevalu F. Ofahengaue Vakalahi** Morgan State University School of Social Work, Baltimore, MD, USA

**Ricardo Pérez-Cuevas** National Institute of Public Health, Cuernavaca, Morelos, Mexico

**Mario Ulises Pérez-Zepeda** Department of Geriatric Epidemiology, National Institute of Geriatrics, Mexico City, Mexico

**Mario Enrique Rendón-Macías** Research Unit in Clinical Epidemiology, Pediatrics Hospital, National Medical Center Century XXI, Mexican Institute of Social Security, Mexico City, Mexico

**Hortensia Reyes-Morales** National Institute of Public Health, Cuernavaca, Morelos, Mexico

**Sergio Sánchez-García** Chief, Research Unit in Epidemiology and Health Services, Aging Area, National Medical Center Century XXI, Mexican Institute of Social Security, Mexico City, Mexico

**Laurens G. Van Sluytman** Morgan State University School of Social Work, Baltimore, MD, USA

**Pamela Tella-Vega** Department of Geriatric Epidemiology, National Institute of Geriatrics, Mexico City, Mexico

**Carmen Jimena Vázquez-García** University of Essex, Colchester, UK

**Ernesto Velasco-Sánchez** CIVICUS Consultants, Mexico City, Mexico

**Laura Bárbara Velázquez-Olmedo** Faculty of Dentistry, National Autonomous University of Mexico, Mexico City, Mexico

**Miguel Ángel Villasís-Keever** Research Unit in Clinical Epidemiology, Pediatrics Hospital, National Medical Center Century XXI, Mexican Institute of Social Security, Mexico City, Mexico

**Fernando A. Wagner** University of Maryland School of Social Work, Baltimore, MD, USA

**Chioma Nwakanma Wosu** University of Maryland School of Social Work, Baltimore, MD, USA

**Leonardo Zapata-Fonseca** Student, Program of the Plan of Combined Studies in Medicine, MD/PhD, Faculty of Medicine, National Autonomous University of Mexico, Mexico City, Mexico

**Elisa Zengarini** Department of Medicine, Geriatric Institute University of Perugia Medical Faculty, Perugia, Italy

**Tirso Zúñiga-Santamaría** Neurogenetics Department, National Institute of Neurology, Mexico City, Mexico

# Chapter 1
# The Need for Differentiated Research Methodology in Aging

**Mario Ulises Pérez-Zepeda, Carmen García-Peña, and Luis Miguel Gutiérrez-Robledo**

**Abstract** The global phenomenon of population aging has increased the need of accurate information in the last few years. In order to improve current status of the older adult's care, quality information should be generated by standardized research methodology. Specific issues arise when it comes to aging research, different from those found in younger stages of life. Having this in mind and how could impact the older adult will result in a continuous generation of helpful information for evidence-based decision making in all levels of older adult care.

**Keywords** Aging · Geriatric research · Evidence-based geriatrics · Older adult care

## 1.1 Introduction

It is well known that in the years to come the population group with the highest growth will be that of the elderly and that this in turn brings with it a specific demand for health care and in all areas of human activity [1, 2]. Health sciences in particular face a major challenge to maintain the well-being of older adults, since it is at this stage of life where to draw on all disciplines, it is essential to be successful. That is, it is not enough to generate information from a single point of view, it is necessary to integrate the components of a phenomenon into a single vision, for example: diabetes mellitus from the perspective of molecular biology can shed light on the phenomena that occur in this level to give rise to disability in an older adult, but it is also necessary to know what real impact this disease has at the epidemiological level, to know the pharmacology of the different medicines, to know the social influence in the health disease process, to know what elements of technology could

M. U. Pérez-Zepeda (✉)
Department of Geriatric Epidemiology, National Institute of Geriatrics, Mexico City, Mexico
e-mail: mperez@inger.gob.mx

C. García-Peña · L. M. Gutiérrez-Robledo
National Institute of Geriatrics, Mexico City, Mexico
e-mail: mcgarcia@inger.gob.mx; lmgutierrez@inger.gob.mx

 
C. García-Peña et al. (eds.), *Aging Research - Methodological Issues*,
https://doi.org/10.1007/978-3-319-95387-8_1

improve the overall health status of diabetic older adults and once this information is obtained integrate it to establish action plans for diabetic older adults. This is why it is necessary to have solid data that allows to have accurate strategies in the shortest time to achieve this end. The aging process is not synonymous with a decrease in the function of the human body (or any other living organism), however, it does bring with it an increase in the frequency of chronic diseases, mainly diabetes mellitus and systemic hypertension [3]. Today, one of the greatest health challenges of this age group is knowing how to age with a chronic disease, what are the effects of the complications of the diseases themselves, what impact does the long-term use of medications have, it is the role of chronic diseases in the loss of functionality; among many other topics in which there is still not enough information to carry out concrete actions [4]. The most common outcome in recent years has been simply to carry out the same actions that are used in younger adults, a strategy that is not effective and in the worst case has been harmful. There are some examples in elderly Mexicans, where it has been shown that implementing a specific care strategy for older adults can improve their health status [5].

Disability and the so-called geriatric syndromes is another field with a great knowledge gap. As previously mentioned, knowing the different perspectives of these phenomena helps to have global solutions and with less margin of error when taking into account the elderly with their bio-psycho-social environment and with less emphasis on the "organicist" vision that Currently prevails in medicine, in other words, what is good for a kidney is not necessarily good for the heart.

There are two problems of particular attention in the health of older adults: dementia and frailty. Dementia is better known today and many of the resources in research are currently being devoted to its study, however it remains a condition with a high burden for those who have it and particularly for their family and social environment [6]. On the other hand, frailty - understood as the loss of the ability to respond to harmful stimuli - is still an emerging problem and with many questions still to be resolved [7].

The research that nursing has provided for aging has been spearheaded in many ways, just to mention an example, the main interventions in dementia available today are interventions designed, tested and tested within the context of research in care nursing [8]. The research potential in this field is great, however the continuous challenge is to establish links and articulate with other disciplines to perfect the knowledge acquired when the research is done from a unique and isolated perspective.

In addition, it is important to mention that the WHO is trying to shift the paradigm from disability to healthy aging, in order to tackle progression from a starting point, rather than it's too late.

## 1.2 A Theoretical Frame for Aging Research

Different disciplines are grounded in theories that give the topic a sense and a cohesion that could further be enriched by new knowledge. Aging research has a number of potential theoretical frameworks that could be used for this purpose. However, as it happens in other medical disciplines, there is no agreement on which particular theory is appropriate. Evolution of species is one of the most useful theories to explain aging and has fully translated into aging phenomenon by the disposable soma theory of Kirkwood. Moreover, many of the processes that occur during aging seem to respond to evolution.

## 1.3 Particular Features of Aging Research

Once the aging of the population and the lack of information on this age group is recognized as a problem, the question arises about what distinguishes scientific research on health in this area from other disciplines [9].

Research into age and aging uses the scientific method to generate knowledge, not unlike other disciplines. However, it incorporates many more elements than those generally used in "traditional" health research. One of the main differences is the focus on the preservation of function of older adults at different levels (with difficulty, independent, dependent, etc.), in contrast to the objective of preserving life, which is more usual in health research in other age groups [9]. On the other hand, the reductionist focus of other medical specialties (internal medicine, surgery, orthopedics, etc.) makes it difficult to study the phenomenon of aging and it is more useful, both conceptually and in practice, to focus on the biology of systems, or a holistic approach [10]. Another type of focus that can be useful is called "subject-centered", in which the weight of the signs of discomfort by the persons involved acquires more relevancy than the numbers from biochemical measurement [11].

The incorporation of more topics of investigation than is the case at present will be done in the years to come. Among the new items to consider are: services (access, quality, innovation, technology), the incorporation of social determinants of health, deep analysis of these determinants, a multi-disciplinary approach, systematic incorporation of the evidence for creation of public policies, and molecular biology (genomics, proteomics, metabolomics) [9]. As well, in a world of limited resources, research in the economics of health is a fundamental ingredient for the creation of knowledge for improving the clinical care of older adults. Those changes (if any) will have to be adapted to the group of older adults.

Research into age and aging is no different from other research; it simply has emphasized some characteristics that are often harder to investigate in this age group, such as: defining what is normal (normal changes in aging *vs.* pathological changes), "normalization" of problems/illnesses of age, nihilism (thinking that whether or not something is done, why do research in this age group if they will soon die or be incapacitated?), non-specific manifestations of problems, coupled with homogenous definitions – bias in classification – (the case of frailty, whose variability shows up in studies of it), the need for adequate sources of information (valid scales, trained interviewers and optimization of obtaining and analyzing data, to name a few) [9, 12].

There are a number of examples on how aging is different from other type of research, in this chapter some of them will be reviewed.

### *1.3.1 Heterogeneity in Older Adults*

With the goal of having a framework of heterogeneity in age, what follows is a description of different groups, very differentiated within this population segment. With the advances in knowledge about aging in recent decades, a group that previously appeared to be homogeneous now is known to be made up of distinct subgroups, whose characteristics must be taken into account in the various domains at the time of doing the research [13]. Even though there is agreement about the age at which a person should start to be called old (older than 60 years), this does not always correlate biologically [14]. There are sub-groups with specific characteristics, whose differences must be taken into account throughout the design and development of any research project into age or aging: sampling (over-sampling of barely representative groups), selection criteria, stratification, allocation of the intervention, statistical adjustments, in a way that real conclusions are arrived at and not derived from population differences established *a priori* (see Table 1.1). Another characteristic that generates different sub-groups, and that it is crucial to take into account, is related to the losses in the trials, since in some cases they are highly characteristic, for example, in subjects with dementia.

Therefore, it is necessary to thoroughly know these different groups within the group of older adults, in order to be able to make the pertinent adjustments in the design of the protocol, or in the last instance, if this is not possible, at least to describe the population group and its distinct characteristics. The following is a detailed description of some of the characteristics that produce the marked differences.

#### 1.3.1.1 Age

The easiest way to look at for this category is chronological: the more years that have passed, the higher the probability of suffering one or more illnesses, and the same with loss of function, frailty and the appearance of geriatric syndromes. Therefore, the division of these groups by age in research has a clinical logic. In addition and depending on the outcome – following the Gompertz curve – it is known that the probability of dying is greater with advancing age, a situation to take into account, for example, when comparing groups of subjects of 60 and 90 years of

**Table 1.1** Different groups to take into account in age and aging research

| Group | Categories |
|---|---|
| Age | Young-old 60 to 79 |
| | Old-old 80 to 89 |
| | Extremely old 90 and above |
| | Nonagenarian 91 to 100 |
| | Super-centenarians older than 101 |
| Function | Effective function without difficulty |
| | Effective function with difficulty |
| | Ineffective function without difficulty |
| | Ineffective function with difficulty |
| | Loss of function in some activities, with dependence with assistance |
| | Loss of function in some activities with dependence without assistance |
| | Loss of function in all activities with dependence with assistance |
| | Loss of function in all activities with dependency without assistance |
| Multi-morbidity/ Polypathology | Without non-degenerative chronic illnesses |
| | With one non-degenerative chronic illness |
| | With multi-morbidity/polypathology |
| Life prognosis | Without terminal illness |
| | With a terminal illness but without probability of dying in the next 6 months |
| | With terminal illness with probability of dying within the next 6 months; not moribund |
| | Moribund |
| Specific pathology | Without a specific pathology |
| | Dementia |
| | Cancer |
| | Frailty |
| Level of care | Ambulatory |
| | Acute hospital care |
| | Chronic hospital care |
| | Residence |
| | Hospice |
| Caregiver | Without a caregiver |
| | Without caregiver burden |
| | With caregiver burden |

age. If one wants to evaluate the effect of a particular intervention, wants to show the impact on mortality, and is unable to find any, the difference by age – expected and not adjusted – would be the explanation [15]. Finally, the group most advanced in age, the people older than 100 years of age, is much less represented in the studies, being one of the most forgotten groups in all types of research.

Taking into account the foregoing, conventionally the most common way to divide groups of older adults by age is: "old-young" 60 to 79; "old-old" 80 to 90; "ancient old-old" 90 and over; "nonagenarians" 91 to 100; and "super centenarians" greater than 101 years [16]. As can be observed, this is an arbitrary division and within each group there is also a lot of heterogeneity, given that health strategies not only provide for an increase in life expectations, but also an increase in the expectation of a healthy life. Alternatives to the division by age groups could be those given by levels of functioning, the extent of non-transmittable chronic illness, specific pathologies (cancer, dementia, etc.), level of health care required or frailty status.

#### 1.3.1.2 Function

Defined as the capacity to be able to carry out, independently and autonomously, the activities necessary to take care of oneself under optimal conditions and within one's own surroundings, function is an effective way to classify the elderly. There is a large spectrum between the two extremes (independence and dependence), with a number of activities within which is also a different range of effectiveness in capacity to carry out these functions (independence in function, difficulty in doing this, and total dependence on someone else to be able to do some of the activities). Taking into account the potential effect function can have on a particular intervention improves the possibility of obtaining other appropriate outcomes. Research can also be carried out on specific groups of levels of functional, as is the case in researching the cause of pressure sores in people almost totally dependent; testing an intervention on injuries to cure them. The fact of not taking into account functional could give the false impression of the functional and the effectiveness of the intervention.

#### 1.3.1.3 Multi-Morbidity

Recently there has been emphasis on this concept (suffering from more than one non-transmittable chronic illness for which the person is taking medicine regularly), because it appears that it could involve a problem with characteristics different from the rest of the population and an entity in which there would have to be special care taken in carrying out clinical tests at the time of reporting the interactions. This is the case not only with the medicines but also with the illnesses, and the potential synergies that might exist. The strategy of excluding competing risks is known, that it is not possible to control them and that they could bring about the outcome they intend to change with the intervention. The foregoing is especially true when dealing with cohort studies in which an exposed factor tries to associate with a specific outcome, since with the passage of time there are other exposures that could bring about the same outcome, for example in the case of falls – which can be caused by multiple factors – and not taking into account several causes for the adjustment in one of the moments of the research project [17]. However, several solutions have been promoted in the area of old people. One of the most common is the use of indices of comorbidity, for example Charlson's, in which the results can be adjusted specifically to this comorbidity burden, and theoretically eliminating this potential source of confusion in the clinical tests. To choose the population *a priori*, in such a way as to exclude some with competing risks would hardly be practical (and hardly realistic) in research in older adults, given the low probability of finding "healthy" old people or people with just one pathology.

### 1.3.2 Animal Models

Even that it has been said that modeling of aging is somewhat a closed matter, and we do not have the need to discuss it further, there have been some problems with modeling animals due to the complex nature of older adults. The typical example is frailty, a multi-level problem that renders an older adult prone to adverse events in the face of usual stressors, which has been shown to be difficult to model in animals.

### 1.3.3 The Role of Time

Time is one of the most difficult issues to handle in aging research. One thing is to have age-related problems, which have been defined as those that are only related with time passing by and the other is that of age-dependent problems in which those mechanisms that render an organism old, also have a role in disease.

### 1.3.4 The Role of Outcomes

Death is more common for the aged ones. In other kind of studies, mortality is the main outcome. However, when it comes to older adults, mortality is not always the best outcome to measure, and other outcomes are more important.

### 1.3.5 Statistical Approaches

A recent number on the journal of gerontology series B has been devoted exclusively to the description of different solutions to a number of problems faced in aging research. This is particularly true when it comes to analyzing longitudinal studies, in which attrition rates are high. Competitive risks are among other type of problems faced when analyzing longitudinal studies and related outcomes in population.

## 1.4 Conclusions

To better understand the problems that could be presented at each stage of the research in areas of age and aging, knowledge is created clearly and with a solid scientific structure that contributes to improvement in the quality of care for older adults and a clearer understanding of the processes that could have impact on their overall health.

## References

1. Ham Chande R, Gutierrez Robledo LM (2007) Health and aging in the 20th century. Salud Publica Mex 49(Suppl 4):S433–S435 doi: S0036-36342007001000001
2. Gutierrez-Robledo LM (2002) Looking at the future of geriatric care in developing countries. J Gerontol A Biol Sci Med Sci 57(3):M162–M167
3. Kinsella KG (2005) Future longevity-demographic concerns and consequences. J Am Geriatr Soc 53(9 Suppl):S299–S303. https://doi.org/10.1111/j.1532-5415.2005.53494.x
4. Pang T, Sadana R, Hanney S, Bhutta ZA, Hyder AA, Simon J (2003) Knowledge for better health: a conceptual framework and foundation for health research systems. Bull World Health Organ 81(11):815–820 doi: S0042-96862003001100008
5. Remme JH, Adam T, Becerra-Posada F, D'Arcangues C, Devlin M, Gardner C et al (2010) Defining research to improve health systems. PLoS Med 7(11):e1001000. https://doi.org/10.1371/journal.pmed.1001000
6. Orton L, Lloyd-Williams F, Taylor-Robinson D, O'Flaherty M, Capewell S (2011) The use of research evidence in public health decision making processes: systematic review. PLoS One 6(7):e21704. https://doi.org/10.1371/journal.pone.0021704
7. Steel K (1997) Research on aging. An agenda for all nations individually and collectively. JAMA 278(16):1374–1375
8. Cordova-Villalobos JA, Barriguete-Melendez JA, Lara-Esqueda A, Barquera S, Rosas-Peralta M, Hernandez-Avila M et al (2008) Chronic non-communicable diseases in Mexico: epidemiologic synopsis and integral prevention. Salud Publica Mex 50(5):419–427 doi: S0036-36342008000500015
9. Dartigues JF (2005) Methodological problems in clinical and epidemiological research on ageing. Rev Epidemiol Sante Publique 53(3):243–249
10. Ahn AC, Tewari M, Poon CS, Phillips RS (2006) The limits of reductionism in medicine: could systems biology offer an alternative? PLoS Med 3(6):e208. https://doi.org/10.1371/journal.pmed.0030208
11. Reuben DB, Tinetti ME (2012) Goal-oriented patient care--an alternative health outcomes paradigm. N Engl J Med 366(9):777–779. https://doi.org/10.1056/NEJMp1113631
12. Van Ness PH, Charpentier PA, Ip EH, Leng X, Murphy TE, Tooze JA et al (2010) Gerontologic biostatistics: the statistical challenges of clinical research with older study participants. J Am Geriatr Soc 58(7):1386–1392. https://doi.org/10.1111/j.1532-5415.2010.02926.x
13. Friedman LM, Furberg C, DeMets DL (2010) Fundamentals of clinical trials, 4th edn. Springer, New York

14. World Assembly on Aging (1983) A neglected area in the field of population and human rights: aging and the aged. United Nations, New York
15. Winsor CP (1932) The Gompertz curve as a growth curve. Proc Natl Acad Sci U S A 18(1):1–8
16. Orimo H (2006) Reviewing the definition of elderly. Nihon Ronen Igakkai Zasshi 43(1):27–34
17. Rothman KJ, Greenland S, Lash TL (2008) Modern epidemiology, 3rd edn. Wolters Kluwer Health/Lippincott Williams & Wilkins, Philadelphia

# Chapter 2
# The Scientific Method as a Point of Departure in Aging Research

**Rubén Fossion and Leonardo Zapata-Fonseca**

**Abstract** What makes knowledge scientific is not its content per se but rather the form, in which it is obtained. Following the scientific method is a necessary condition to carry out a sound and methodologically valid research. However, for empirical researchers, it is not common practice to reflect upon the method itself. It has been argued that the scientific method is not so different from the common sense that we use in daily life to reach solutions, but with its successive steps better articulated so that scientific knowledge can approach more robust conclusions over time. Since the last quarter of the previous century, there are indications that reductionist strategy of the scientific method has reached its limits, and that therefore a complementary approach is needed to investigate new complex research problems. Consequently, emergentism and systemic thinking are becoming a new explanatory framework that is currently permeating virtually any field of knowledge and all spatiotemporal scales. In the present chapter, we focus on a very specific system under a rather specific yet common and relevant condition: the aging human being. Particularly, we introduce some notions on how the sciences of complexity can help, not only clinicians but also medical research in general –and in particular aging research– to reach a more complete understanding and assessment of the older adult both at an individual and population levels.

**Keywords** Philosophy of science · Reductionism · Complexity · Effective theory

R. Fossion (✉)
Nuclear Sciences Institute and Centre for Complexity Science (C3), National Autonomous University of Mexico, Mexico City, Mexico
e-mail: ruben.fossion@nucleares.unam.mx

L. Zapata-Fonseca
Faculty of Medicine, National Autonomous University of Mexico, Mexico City, Mexico

 
C. García-Peña et al. (eds.), *Aging Research - Methodological Issues*,
https://doi.org/10.1007/978-3-319-95387-8_2

## 2.1 Introduction: History and Philosophy of Science

Society esteems science for its presumed quality of being based on objective facts, so that scientific research has more weight and authority than a personal opinion [1]. But what makes something scientific? It is not the object or the topic under study but rather the methodology with which a study is carried out and the standards that are used to judge the obtained results [2]. The methodology that is used in science, or the so-called scientific method, is not very different from the way in which we use common sense to interpret events in our daily lives. Common sense analyzes the information we receive through our senses (sight, hearing, touch, smell, taste) as being real and independent from the observer. Without thinking consciously about the steps taken, our common sense is based on a sequence of observation, evidence and verification; scientific thinking follows the same logic, but the scientific train of thought is slowed down for increasing transparency and control during the various steps. Transparency is important because scientific research is a collaborative activity and peers and colleagues must be allowed to repeat experiments, verify results and construct more advanced theories based on previous results [2].

The scientific method developed gradually through several millennia. The Ancient Greeks, such as Socrates (469–399 BC), Plato (427–348 BC), Aristotle (384-322 BC), Ptolemy (90–180 AD) and Galen (130–200 AD), were pioneers in setting up a science independent of religious dogmas; theirs was mostly a contemplative science based on abstract axioms to which they applied deductive logic in order to obtain new statements, at most passive observations of Nature were made and induction was used to obtain new hypotheses. During the dark ages of Europe's Medieval Period (500–1300 AD), much of the scientific knowledge of the Ancients Greeks was lost. Fortunately, a lot of that knowledge could be recovered thanks to the Arabs (700–1500 AD) who had adopted the science of the Ancient Greeks and who had contributed with active experimentation, which was an important step forward because now theoretical predictions were verified with experiments. The next important period is the scientific revolution (1500–1800 AD), which was caused by various factors. One factor was the foundation of the first universities, which resulted in a gradual "liberalization" of the sciences and leading to a more pluralistic vision not dictated by a few authorities. Another factor was humanism as the new philosophical and ethical current having as one of its purposes to explain all-natural phenomena without any reference to the supernatural. Technological inventions, such as the microscope and the telescope, further accelerated the advance of science. Also, important resulted to be mathematical modeling, which allows researchers to make not only qualitative but also quantitative predictions.

The study of the history of the scientific method and how the scientific method is applied, is a science of science, also called meta-science, and therefore belongs to the field of philosophy of science. Philosophy is a forum to question and clarify concepts that other disciplines believe to be obvious without having investigated these questions explicitly [2]. Philosophy of science analyzes the various steps of a scientific investigation. Consequently, the philosophical approach tends to be abstract and idealistic, and the goal is to define an absolute and universal scientific method that is valid for all disciplines and for all times.

In the application of the scientific method, the following properties are often taken for granted:

- The data are previous to and independent from theory;
- The data constitute a firm and reliable base for scientific knowledge;
- The experimental data are obtained by impartial observation through the senses.

Philosophers have identified some problems with these assertions, such as theory and subjectivity ladeness [3], confirmation and rejection of the theories [4, 5], and how to evaluate scientific progress [1].

The American physicist and philosopher of science Thomas Kuhn (1922–1996 AD) revolutionized the way in which scientific progress is perceived. Before Kuhn, scientific progress was interpreted as a gradual process; it has been suggested that our textbooks are to blame for reinforcing this view of a continuous accumulation of ideas up to the current state of science, while Kuhn argues that scientific achievements of the past need to be interpreted within the context of sociological factors and scientific perspectives of the time in which they were developed [6]. It appears that within each scientific specialty, prolonged periods of stability and consolidation precede short bursts of major conceptual revision, which Kuhn called paradigm shifts [7]. A paradigm is a coherent set of theories and concepts that guides interpretations, the choice of relevant experiments, and the development of additional theories in a field of study. Examples of contrasting paradigms in physics are: heliocentrism vs. geocentrism, Newtonian gravity as opposed to Einstein's theory of general relativity, and classical physics versus quantum mechanics. In medicine, examples of paradigm shifts are the dissection of human cadavers as introduced by Vesalius, the use of the microscope and the development of synthetic drugs.

Standard science works within the framework of an existing paradigm that guides a field of research. In this case, almost all the research relates to the paradigm: research is carried out according to a fixed scheme, and it is the paradigm that indicates which topics for research are appropriate and worthwhile; theoretical and experimental studies imply the collection of data to verify predictions of the paradigm and consider also efforts to extend the paradigm in order to include apparent problems or ambiguities. Research within an existing paradigm is sometimes described in a pejorative way as "cleaning up". In a new field, that is, a field in a pre-paradigm state, no fixed scheme exists that indicates how experiments should be done or how data should be interpreted. To draw an analogy: data collection within the framework of an existing paradigm is like a hunter pursuing a prey, while without the guidance of a paradigm it rather resembles going for fishing in a lake to see what comes out [6]. In the absence of a paradigm, lots of data may be available but they are extremely complicated to interpret, and the general pattern and the main principles are vague; several currents of reasoning compete without agreement on which phenomena are worth studying, and no single current of reasoning can offer a more general view of the field.

## 2.2 A Pragmatic Approach to the Scientific Method

In comparison with philosophers, working scientists are more realistic and conformist, and are satisfied with an approximated scientific method that in the first place must be applicable to their daily research activities. The structure of the scientific method, in its most basic form, can be summarized as successive repetitions of the following sequence, see Fig. 2.1:

*Observation→Taxonomy→Working Hypothesis→Prediction→Empirical Verification*

In the observation phase, relevant data about a natural phenomenon of interest are recognized. The taxonomy stage detects and classifies regular patterns in the data. The induction[i] phase enables the researcher to generalize and simplify these patterns in one or more theoretical hypotheses to explain the phenomenon. Abduction[ii] is a type of logical inference that is used to select the most probable hypothesis from a set of possible hypotheses to explain a given phenomenon. Applying deductive[iii] logic to the working hypothesis allows to derive predictions, which can be verified with the results of carefully controlled experiments. A controlled experiment is one where a certain (independent) variable is manipulated to study the consequent changes in another (dependent) variable. It is preferable that

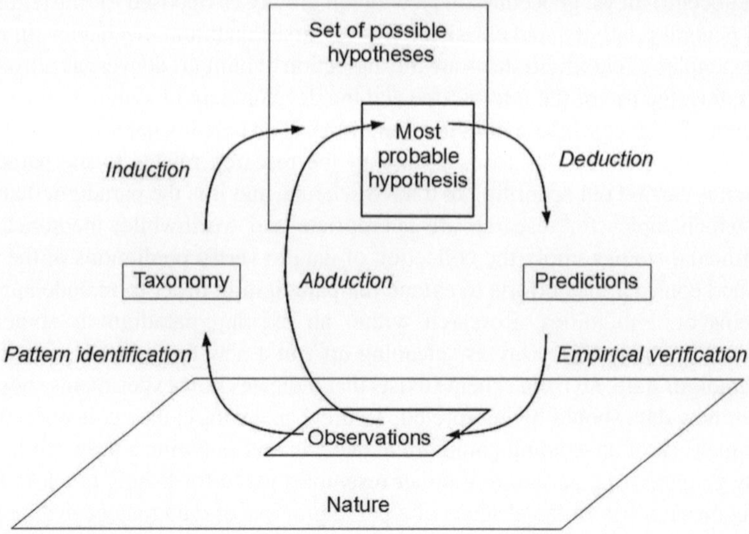


**Fig. 2.1** The process of scientific reasoning is iterative and alternates between deduction, induction and abduction. Induction generalizes observed patterns in nature in theoretical models. Abduction selects the most probable working hypothesis from a set of hypotheses to explain an observed phenomenon. Using deduction, predictions are made from the hypothesis to be verified with data from controlled experiments, so that the hypothesis can be checked and corrected if necessary

all the other variables stay constant to avoid confounding factors. When new observations are made, and new experimental data are obtained, the working hypothesis may be retained, modified or refuted. It is assumed that the repetition of the sequence … → Observation → Taxonomy → New Hypothesis → Prediction → Experimental Verification → ... will converge to an accurate description of the true state of Nature [6, 8–10].

According to the great German quantum physicist Werner Heisenberg (1901–1976) science does not provide an objective explanation of Nature; rather, it describes what is exposed of Nature through the specific method of questioning being used [11]. This results in the following paradox: how is it possible to reconcile the apparent and profound success of science with the problem that scientific objectivity might be an elusive ideal because of the inherent subjectivity in perception? Science depends less on absolute objectivity than is thought traditionally. It can be argued that scientists are in the first-place pragmatists: the challenge is to use methods and assumptions, bearing in mind that they are subjective and imperfect, and at the same time try to obtain an as objective as possible understanding of the patterns and principles of Nature [6]. Certainly, any scientific investigation must necessarily use a biased scale to weigh and evaluate data because all scales are biased; but if we are fully conscious of the bias in the scale, it can be used effectively. To increase the precision of the scale, we need to know the sources of error. To achieve this, we need to understand the limits of our methods and it is important to understand, too, how the process of perception affects our observations and thus be able to recognize our own biases.

A first step toward the recognition of biases in research is complete transparency, or full disclosure, to reveal all the elements and steps taken to arrive at a scientific conclusion. Already Aristotle was interested in the transparency of scientific reasoning: "What is it that goes in so that scientific conclusions come out?" he asked. A modern model of transparency is the PEL model [9], see Fig. 2.2 The first element of the PEL model is the list of presuppositions (P), which offers a basic and indispensable image of the system being studied. The presuppositions are important because they enable one to restrict the set of all possible hypotheses, which are infinite, to a limited set. Without constraint, the set of possible hypotheses would be infinite and it would be impossible to reject all the absurd hypotheses and keep the realistic ones, based on the finite quantity of empirical evidence accessible to us. Presuppositions do not differentiate between the credibility of each of the realistic hypotheses because the presuppositions are what all hypotheses have in common. On the other hand, evidence (E) is data that can distinguish among the different hypotheses. Finally, logic (L) combines the premises of presuppositions and evidence with logical reasoning (deduction, induction and abduction) to arrive at a conclusion. Importantly, it must be always acknowledged the ubiquitous uncertainty that exists even in the most exact descriptions.

The scientific method and its companion from daily life, common sense, are always confronted with uncertainty being sensitive to levels of certainty: everything exists on a spectrum between mere conjecture and absolute certainty. The

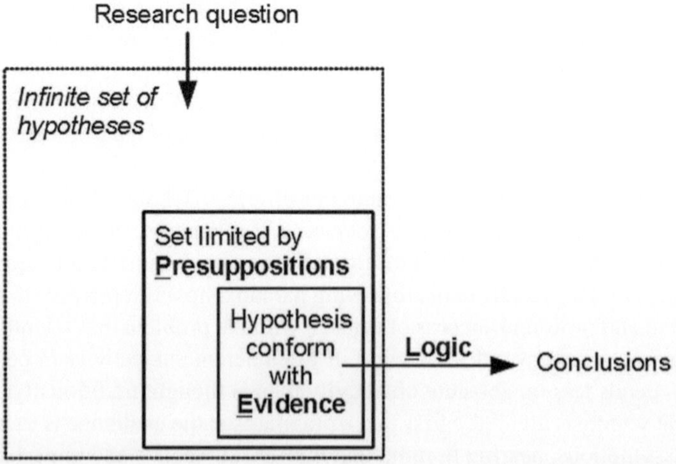


**Fig. 2.2** Complete transparency, or full disclosure, of a research project according to the PEL model. The set of all hypotheses is infinite. The presuppositions (P) enable to discard unrealistic hypotheses. The evidence (E) allows to choose a proper working hypothesis. Logic (L) combines the presuppositions and the evidence to arrive at a conclusion

aim of the scientific method is to localize a theory within this spectrum; when evidence accumulates, the position of the theory on the spectrum can change. To increase our confidence in a hypothesis large numbers and varieties of different tests are necessary. One good way to gain confidence in a hypothesis is to verify how it fits within a coherent network of other theoretical statements and experimental claims. The concept of a coherent network of ideas plays an important role within the scientific method. The most important requisite for coherency is logical consistency because a network of knowledge does not tolerate contradiction. This does not mean that inconsistencies that persist in diverse scientific specialties cannot exist, but when such inconsistencies are found they need to be studied and, in the end, usually give way to new scientific discoveries; that is why contradictions in science cannot be ignored. In addition to being consistent, scientific claims should be cooperative, which means that they need to generate connections between different ideas, such that a new claim in one field of science can play the role of an auxiliary theory in the same or in other scientific disciplines. There is a variety of links between different theories, resulting in an interrelated and coherent network of scientific claims. When evidence accumulates and a new theory becomes better interconnected with other theories, the classification of this theory as uncertain slowly disappears and the theory converges toward the side of certainty on the spectrum. When a hypothesis arrives at equilibrium within a network of scientific knowledge, confidence in its certainty is established; this is also part of the scientific method [2, 12].

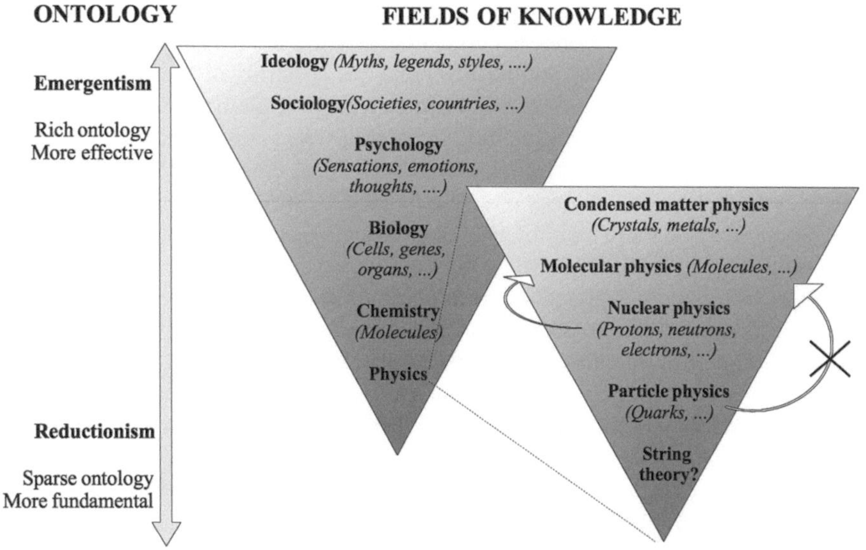


**Fig. 2.3** Different fields of knowledge handle different ontologies or "kits of building blocks". Fields that are "higher up" tend to have richer ontologies with a large variety of building blocks. At each level, these building blocks can effectively be treated to be basic, although they emerge from the sparser and more fundamental building blocks from the field "directly below". Instead, fields are shielded from levels "further down". Philosophy of science, as discussed in Sect. 2.1, is a meta-science, which transcends the knowledge fields mentioned here, and which can be applied to investigate the validity of the scientific method followed in each of the specific fields

## 2.3 The Scientific Method in the Modern Context of Complex Systems and Applied to the Complex Problem of Human Aging

One of the oldest research questions of humankind has been to explain the immense diversity of objects and phenomena met in daily experience. Different strategies exist to investigate this ontology, or basic view of what exists in the world [12]. Reductionism is one strategy, where a richer ontology at one level is explained by a sparser ontology at a more fundamental level, see Fig. 2.3 As an example, all chemical elements of Mendeleev's periodic table, from atomic number 1 (hydrogen H) to 118 (oganesson Og), can be explained as different combinations of only 3 basic building blocks: protons, neutrons and electrons, constitutes by the even more elementary quarks.

The complementary approach to ontology is emergentism, where one tries to understand how basic building blocks and the interactions between them may generate collective patterns of order at a higher level. As an example, all water consists of $H_2O$ water molecules, but in order to navigate at sea, it makes more sense to think of waves as basic components for the dynamics instead. The reason many independent fields of knowledge have developed, from sociology and psychology to biology and chemistry, instead of reasoning from the most elementary building blocks of matter that have been confirmed experimentally (which are the quarks from particle physics), is that no field is completely reducible in terms of more fundamental fields. In other words, new properties arise or emerge from the collective behavior of more basic building blocks and their interactions, and these emerging properties cannot be predicted from first principles [13, 14]. These emerging collective properties may form effective degrees of freedom, such as protons and neutrons, or sea waves, and may be basic building blocks for practical purposes within the limits of the corresponding level. Effective theories built upon these effective degrees of freedom are shielded from effects from more elementary levels, which implies that effective theories stay true forever, even when profound new scientific advances are made. More general theories (such as Einstein's general relativity theory) incorporate previous theories (such as Newtonian gravity) as a special case, but these older theories remain valid within the boundaries of applicability for which they were originally proposed [12].

One of the most important collective properties is resilience, which is not so easy to understand and even more difficult to quantify; it is the antonym of frailty, and it can loosely be interpreted as the combination of robustness and adaptability to withstand and/or adjust to perturbations from the internal and/or external environment [15–20]. The main underlying hypothesis of complexity science is that emergence and collectivity in many different fields of knowledge share common properties and may be described in a unified way, such that methods borrowed from one discipline may be used to understand completely different systems in a completely different discipline operating at completely different scales; obviously, one should exert caution when importing from and exporting to other disciplines as the similarity is at the level of the tools and should not be assumed to have any validity beyond that. Therefore, one of the great benefits of complexity thinking is that it opens up a new interaction between the social and natural sciences and provides a language in which the two can communicate [21]. There have been previous attempts towards a transdisciplinary approach to science, such as cybernetics [22, 23], general systems theory [24] and catastrophe theory [25]; and one of the open questions is whether the study of complex dynamical systems will establish itself as a branch of science or whether its methods and techniques will be absorbed into the different disciplines of mainstream science [21].

Since the 1970s an epidemiological transition has been observed, associated to the aging of populations in many countries all over the world, from a predominance of acute infectious diseases to a higher prevalence of chronic-degenerative illnesses [26]. Acute diseases (e.g., a bacterial infection or a bone fracture) are usually relatively "simple" to diagnose and treat because often it is possible to localize and delimit the affected part of the body, and although several risk factors can be in play, the causes of the symptoms are quite clear in general. In contrast, chronic-degenerative illnesses like cancer, diabetes, chronic stress, fibromyalgia, etc. seem to be more "complex". Those afflictions are usually systemic, whereby several organs or biological processes are affected simultaneously, and multifactorial, with a broad spectrum of risk factors ranging from the microscopic (e.g., genetic predisposition) and mesoscopic (e.g., lifestyle) to the macroscopic (e.g., environment). Often it is impossible to find a clear cause-effect relationship, possibly due to a complicated interaction among the multiple risk factors. It is possible that the way the medical world is structured, with a focus on "reductionist" specialization, and the way medical research is carried out, with "reductionist" inclusion and exclusion criteria, is not the most suited to deal with the current high prevalence of "complex" diseases [27, 28]. Human aging is another example of a complex problem, and the difficulties already arise in defining which are the aspects of aging that one wants to study, because different research questions will most probably require different research methods: for example, why do we live as long as we do (longevity); why do we grow old (functional decline); why do we die (mortality) [29]? Should aging be considered as a pathology or as a natural process [30, 31]? Is aging genetically programmed and does it have a purpose or is it a probabilistic process corresponding to a random accumulation of molecular and/or cellular defects [32, 33]? How can one measure and quantify aging objectively, and what is the difference between chronological age and biological age [29]?

One of the reasons for the many ambiguities in aging research mentioned above may be that aging is an emerging property that arises at the level of individual organs, systems of organs, or at the level of the whole organism from the accumulation of molecular and cellular damage [34–37]. Consequently, single biomarkers are improbable to trace the many different aspects affected by the aging process, and instead system-level measures need to be defined that stand for the new effective degrees of freedom that drive aging. An example of such system-level measures are the various frailty scales that have been proposed in recent years, which integrate aspects from physical functionality [38] or from cognitive, emotional and social functionality and the medical history of the older adult [39], and that successfully predict negative age-associated health outcomes. A drawback of these frailty scales is that they are symptom-based, such that they only start to be applicable when the older adult already has lost part of his/her functionality. Therefore, it is of interest to explore whether more fundamental ontological levels, that precede the functionality level, such as the level of physiological regulation, allow to assess age-associated frailty at a pre-symptomatic stage [40], see Fig. 2.4.

Physiological regulation by the autonomous nervous system has been studied in relation to frailty using, e.g., heart rate variability [41] and standardized autonomic reflex tests [42]. Other medical disciplines, such as intensive care and palliative care, struggle with the similar problem of how to define clinical prognostic scales, where vital signs tend to be used for short-term prognosis (hours-days), physical functionality for longer-term prognosis (weeks-months), and where more recently physiological measures such as heart rate variability have been incorporated to successfully improve predictability [43, 44]. Variability of physiological variables can be quantified using complexity measures. The loss of complexity hypothesis states that physiological variables, such as heart rate, lose variability and complexity with aging and disease [45–48]. On the other hand, the hypothesis of early-warning signals proposes that increased variability may reflect that the system is becoming unstable and may collapse [16–20], and high blood pressure variability has indeed

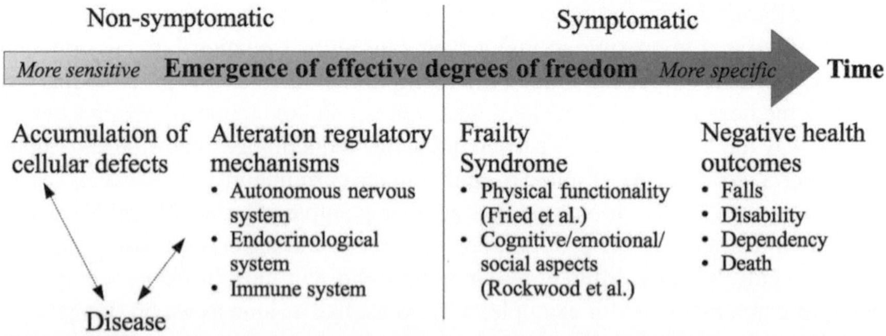


**Fig. 2.4** Aging and age-associated frailty as an emerging property. Clinical prognostic scales for age-associated frailty allow to predict negative health outcomes but are based on specific symptoms and/or functional decline. Symptoms are new effective degrees of freedom that emerge from alterations in physiological regulatory mechanisms that are more sensible but less specific and that could be detected in a pre-symptomatic stage. These alterations themselves may emerge from the accumulation of cellular defects

been interpreted as a risk factor for health [49]. This apparent contradiction may be understood within the context of homeostasis [50], where certain "regulating" physiological variables (e.g., heart rate, respiration dynamics, skin temperature, insulin, etc.) are responsible for the adaptive response to perturbations from the outer environment to ensure the approximate constancy of other "regulated" physiological variables around a specific set point (e.g., blood pressure at 120/80 mmHg, blood oxygen saturation at 95%, core temperature at 37 °C, blood glucose concentration at 100 mg/dL, etc.) which represents the stability of the inner environment. In optimal conditions of youth and health, regulated variables can be expected to vary minimally around a specific set point, whereas regulating variables are characterized by a large variability because of their large adaptive capacity; in adverse conditions of aging and/or chronic-degenerative disease, regulating variables lose their adaptive capacity which is reflected by a loss of variability of the corresponding time series and consequently regulated variables get out of control and the variability around their respective set point increases [51–53].

## 2.4 Conclusions

It is not the *topic* which determines whether something under study is scientific or not, but rather the way in which it is studied; in other words, whether the study follows the scientific method. The scientific method is similar to the common sense that we use in daily life but with the sequence of the different successive steps *Observation* → *Taxonomy* → *Hypothesis* → *Prediction* → *Verification* well-articulated and well documented. Whereas philosophers are idealist and try to define an absolute and universal method, scientists are realists and conformists and are satisfied with an approximate and pragmatic approach that can be applied in daily practice.

Within the scientific method, two complementary strategies exist, which are reductionism and emergentism. The *reductionist* approach is very successful in explaining the wide variety of objects and phenomena (a rich and complicated ontology) in terms of only a few simple building blocks and the interactions between them (a sparse and simple ontology), e.g., all objects of our daily life can be explained as different combinations of only protons, neutrons and electrons. Emergence goes the other way around, asking the question whether we can explain unexpected complex patterns of order given a set of specific building blocks and interactions.

Recently, it has been argued that human ageing and frailty may be emergent phenomena, resulting in yet unexplained ways from molecular and cellular defects that accumulate over time. If this is true, then it may be important exchange methods and techniques with other fields of knowledge, such as economy, climate science and ecology, where the same basic research questions are being investigated of how to deal with emerging complex dynamical systems.

**Notes**

i. *Induction is a tool of logic that simplifies and generalizes patterns observed in a limited amount of data into a theoretical principle (hypothesis, law, model, conjecture). Induction is a creative and imaginative step associated with the inspiration and genius of the researcher. The conclusion does not have absolute certainty, but rather a certain level of probability, which depends on the quality of the evidence. A classic example of induction is observing that all European swans are white and generalizing that all swans must be white. This conclusion was shown to be false when black swans were discovered in Australia.*

ii. *Abduction is a tool of logic that infers a premise from a conclusion. For example, since grass becomes wet when it rains, observing wet grass in the morning, a good working hypothesis might be that it must have rained during the night. Abductive reasoning is prone to the fallacy of affirming the consequent.*

iii. *Deduction is a tool of logic that allows obtaining conclusions from accepted premises. If the first premises are correct, the conclusions are necessarily also correct; in other words, the truth of the premises ensures the truth of the conclusions.*

## References

1. Chalmers AF (1999) What is this thing called science? 3rd edn. Hackett Pub, Indianapolis
2. Kosso P (2011) A summary of scientific method. https://doi.org/10.1007/978-94-007-1614-8
3. Hanson NR (1965) Patterns of discovery: an inquiry into the conceptual foundations of science. CUP Archive, Cambridge

4. Popper KR (1935) Logik der Forschung: zur Erkenntnistheorie der moderner Naturwissenschaft
5. Lakatos I (1969) The problem of inductive logic, proceedings of the international colloquium in the philosophy of science, London, 1965, vol II. North-Holland, Amsterdam
6. Jarrard RD (2001) Scientific methods, an online book. University of Utah, Salt Lake City Google Scholar
7. Kuhn TS (1970) The structure of scientific revolutions. University of Chicago Press, Chicago, pp 84–85
8. Box GE, Hunter JS, Hunter WG (2005) Statistics for experimenters: design, innovation, and discovery. Wiley Interscience, New York
9. Gauch HG (2003) Scientific method in practice. Cambridge University Press, Cambridge
10. McComas WF (2006) The nature of science in science education: rationales and strategies. Springer Science & Business Media, Berlin
11. Heisenberg W (1962) Physics and philosophy: the revolution in modern science [1958]; rpt. Harper & Row, New York
12. Carroll SM (2016) The big picture: on the origins of life, meaning, and the universe itself. Dutton est. 1852, an imprint of Penguin Random House LLC, New York
13. Anderson PW (1972) More is different. Science 177:393–396. https://doi.org/10.1126/science.177.4047.393
14. Anderson PW (2011) More and different: notes from a thoughtful Curmudgeon. World Scientific, Hackensack
15. Zolli A, Healy AM (2013) Resilience: why things bounce back. Simon and Schuster, New York
16. Carpenter SR, Cole JJ, Pace ML, Batt R, Brock WA, Cline T, Coloso J, Hodgson JR, Kitchell JF, Seekell DA, Smith L, Weidel B (2011) Early warnings of regime shifts: a whole-ecosystem experiment. Science 332:1079–1082. https://doi.org/10.1126/science.1203672
17. Scheffer M, Carpenter S, Foley JA, Folke C, Walker B (2001) Catastrophic shifts in ecosystems. Nature 413:591–596. https://doi.org/10.1038/35098000
18. Scheffer M, Bascompte J, Brock WA, Brovkin V, Carpenter SR, Dakos V, Held H, van Nes EH, Rietkerk M, Sugihara G (2009) Early-warning signals for critical transitions. Nature 461:53–59. https://doi.org/10.1038/nature08227
19. Scheffer M, Carpenter SR, Lenton TM, Bascompte J, Brock W, Dakos V, van de Koppel J, van de Leemput IA, Levin SA, van Nes EH, Pascual M, Vandermeer J (2012) Anticipating critical transitions. Science 338:344–348. https://doi.org/10.1126/science.1225244
20. Scheffer M (2009) Critical transitions in nature and society. Princeton University Press, Princeton
21. Gershenson C (2008) Complexity: 5 questions
22. Wiener N (1961) Cybernetics or control and communication in the animal and the machine. MIT press, Cambridge
23. Ashby WR (1957) An introduction to cybernetics. 2nd edn. Chapman & Hall Ltd, London
24. Bertalanffy LV (1968) General system theory, Foundations, Development, Applications. George Braziller, New York
25. Thom R (1975) Structural stability and morphogenesis: an outline of a general theory of models (trans. Fowler DH). Benjamin, Reading
26. Omran AR (2005) The epidemiologic transition: a theory of the epidemiology of population change. 1971. Milbank Q 83:731–757. https://doi.org/10.1111/j.1468-0009.2005.00398.x
27. Ahn AC, Tewari M, Poon C-S, Phillips RS (2006) The limits of reductionism in medicine: could systems biology offer an alternative? PLoS Med 3:e208. https://doi.org/10.1371/journal.pmed.0030208
28. Ahn AC, Tewari M, Poon C-S, Phillips RS (2006) The clinical applications of a systems approach. PLoS Med 3:e209. https://doi.org/10.1371/journal.pmed.0030209
29. Hayflick L (1994) How and why we age. Ballantine Books, New York
30. Bulterijs S, Hull RS, Björk VC, Roy AG (2015) It is time to classify biological aging as a disease. Front Genet 6:205

31. Hayflick L (2007) Biological aging is no longer an unsolved problem. Ann N Y Acad Sci 1100:1–13. https://doi.org/10.1196/annals.1395.001
32. Medawar PB (1957) Uniqueness of the individual. Methuen, London
33. Kirkwood TBL (2005) Understanding the odd science of aging. Cell 120:437–447. https://doi.org/10.1016/j.cell.2005.01.027
34. Mossman KL (2014) The complexity paradox: the more answers we find, the more questions we have. Oxford University Press, New York
35. Cohen AA (2016) Complex systems dynamics in aging: new evidence, continuing questions. Biogerontology 17:205–220. https://doi.org/10.1007/s10522-015-9584-x
36. Mitnitski AB, Rutenberg AD, Farrell S, Rockwood K (2017) Aging, frailty and complex networks. Biogerontology 18:433–446. https://doi.org/10.1007/s10522-017-9684-x
37. Rutenberg AD, Mitnitski AB, Farrell SG, Rockwood K (2017) Unifying aging and frailty through complex dynamical networks. Exp Gerontol. https://doi.org/10.1016/j.exger.2017.08.027
38. Fried LP, Tangen CM, Walston J, Newman AB, Hirsch C, Gottdiener J, Seeman T, Tracy R, Kop WJ, Burke G, MA MB, Cardiovascular Health Study Collaborative Research Group (2001) Frailty in older adults: evidence for a phenotype. J Gerontol A Biol Sci Med Sci 56:M146–M156. https://doi.org/10.1093/gerona/56.3.M146
39. Rockwood K, Song X, MacKnight C, Bergman H, Hogan DB, McDowell I, Mitnitski A (2005) A global clinical measure of fitness and frailty in elderly people. CMAJ 173:489–495. https://doi.org/10.1503/cmaj.050051
40. Fried LP, Walston JD, Ferrucci L (2009) Hazzard's geriatric medicine and gerontology. Halter, JB, pp 631–646
41. Parvaneh S, Howe CL, Toosizadeh N, Honarvar B, Slepian MJ, Fain M, Mohler J, Najafi B (2015) Regulation of cardiac autonomic nervous system control across frailty statuses: a systematic review. Gerontology 62:3–15. https://doi.org/10.1159/000431285
42. Romero-Ortuno R, Cogan L, O'shea D, Lawlor BA, Kenny RA (2011) Orthostatic haemodynamics may be impaired in frailty. Age Ageing 40:576–583
43. Kim DH, Kim JA, Choi YS, Kim SH, Lee JY, Kim YE (2010) Heart rate variability and length of survival in hospice cancer patients. J Korean Med Sci 25:1140–1145. https://doi.org/10.3346/jkms.2010.25.8.1140
44. Brandan ME, Ávila MA, Fossion R, Zapata-Fonseca L (2018) Una mirada a la investigación futura en Física Médica en México. In: Torres Labansat M (ed) Hacia dónde va la Física en México? Fondo Cultural Económico (in print)
45. Lipsitz LA (1992) Loss of "complexity" and aging. JAMA 267:1806. https://doi.org/10.1001/jama.1992.03480130122036
46. Goldberger AL (1996) Non-linear dynamics for clinicians: chaos theory, fractals, and complexity at the bedside. Lancet 347:1312–1314
47. Goldberger AL, Amaral LAN, Hausdorff JM, Ivanov PC, Peng CK, Stanley HE (2002) Fractal dynamics in physiology: alterations with disease and aging. Proc Natl Acad Sci U S A 99(Suppl 1):2466–2472. https://doi.org/10.1073/pnas.012579499
48. Goldberger AL (2006) Giles F. filley lecture. complex systems. Proc Am Thorac Soc 3:467–471. https://doi.org/10.1513/pats.200603-028MS
49. Parati G, Ochoa JE, Lombardi C, Bilo G (2013) Assessment and management of blood-pressure variability. Nat Rev Cardiol 10:143–155. https://doi.org/10.1038/nrcardio.2013.1
50. Modell H, Cliff W, Michael J, McFarland J, Wenderoth MP, Wright A (2015) A physiologist's view of homeostasis. Adv Physiol Educ 39:259–266. https://doi.org/10.1152/advan.00107.2015
51. Fossion R, Fossion JPJ, Rivera AL, Lecona OA, Toledo-Roy JC, García-Pelagio KP, García-Iglesias L, Estañol B (2018a) Homeostasis from a time-series perspective: an intuitive inter-

pretation of the variability of physiological variables. In: Olivares-Quiroz L, Resendis-Antonio O (eds) Quantitative models for microscopic to macroscopic biological macromolecules and tissues. Springer International Publishing, Cham, pp 87–109

52. Fossion R, Sáenz-Burrola A, Zapata-Fonseca L (2018b) On the stability and adaptability of human physiology: Gaussians meet heavy-tailed distributions. INTERdisciplina (CEIICH-UNAM), IN PRESS
53. Fossion R, Rivera AL, Estañol B (2018c) Homeostasis from a physicist point of view: what time series of continuous monitoring tell us about physiological regulation, Physiol. Meas., SUBMITTED

# Chapter 3
# Biomedical Research in Aging

**José Mario González-Meljem, Scott Haston, Suchira Gallage, and Andrew J. Innes**

**Abstract** Biomedical research has been instrumental in identifying key molecular and cellular changes that occur throughout the aging process, also known as the Hallmarks of Aging. Notably, these are shared between humans and several other species that have served as models for the study of aging in the laboratory. In this chapter, we discuss current knowledge regarding the significance of hallmarks such as: decay of stem cell function, acquisition of genomic instability, DNA damage, telomere attrition, deregulated nutrient sensing, chronic inflammation and cellular senescence. We further describe current methodological issues, experimental techniques and best practices for the study of each hallmark across different *in vivo* and *in vitro* systems, while also pointing at their limitations. Finally, we provide future perspectives for the improvement of experimental designs in biomedical research of aging.

**Keywords** Stem cells · DNA-damage · mTOR · Inflammation · Epigenetics · Senescence · SASP

J. M. González-Meljem (✉)
Department of Basic Research, National Institute of Geriatrics, Mexico City, Mexico
e-mail: jmgonzalez@inger.gob.mx

S. Haston
Developmental Biology and Cancer Research Programme, UCL Great Ormond Street Institute of Child Health, London, UK
e-mail: scott.haston.13@ucl.ac.uk

S. Gallage
Division of Chronic Inflammation and Cancer, German Cancer Research Center (DKFZ), Heidelberg, Germany
e-mail: s.gallage@dkfz-heidelberg.de

A. J. Innes
MRC London Institute of Medical Sciences, London, UK

Institute of Clinical Sciences, Faculty of Medicine, Imperial College London, London, UK

Centre for Haematology, Faculty of Medicine, Imperial College London, London, UK
e-mail: a.iness@imperial.ac.uk



C. García-Peña et al. (eds.), *Aging Research - Methodological Issues*,
https://doi.org/10.1007/978-3-319-95387-8_3

## 3.1 Introduction

The main goal of biomedical research is to increase elemental knowledge on the functional mechanisms that underlie both normal and pathological life processes, with a particular attention to those affecting human health. Aging is one such process and occurs widely across organisms. It involves a time-dependent decline in function that spans all organizational levels; from biomolecules to cells, tissues and organs. As is it also the case for many other biological phenomena, the fundamental factors that drive aging in humans appear to be shared to a significant extent with other lifeforms such as yeast, nematode worms, fruit flies and rodents. By studying the events affecting lifespan and time-dependent health deterioration in these models, biomedical research has led to the identification of a number of common denominators of the aging process [1]. These Hallmarks of Aging, include stem cell exhaustion, genomic instability, telomere attrition, epigenetic alterations, deregulated nutrient-sensing, chronic inflammation, and cellular senescence. Due to space constraints we are unable to discuss two other important hallmarks of aging: the loss of cellular proteostasis and mitochondrial dysfunction. These however have been extensively reviewed elsewhere [2, 3].

It is expected that improving our understanding of their role in aging will lead to the development of novel care strategies and therapeutics for preventing age-related disease in humans. For each hallmark, we present current knowledge regarding their importance in aging and focus on discussing methodological aspects such as *in vivo* and *in vitro* systems, experimental tools and techniques, as well as best practices for their application in the context aging. Finally, we will also discuss current limitations of these approaches and provide future perspectives for the field.

## 3.2 Stem Cells in Aging

Stem cells (SCs) are characterized by their dual ability to self-renew and differentiate into most, if not all, of a tissue´s specialized cell types [4]. Due to their ability to continually replenish a population of tissue-specific progenitors, SCs are crucial for growth during embryonic development, regeneration following injury and homeostatic tissue maintenance in adult life [5, 6]. Typically, adult SCs are thought to be essentially immortal, although they are susceptible to damage accumulation during life. In order to appropriately function, SCs also require the action of surrounding supporting cells which are collectively known as the SC niche [7].

As SCs sit atop the hierarchy of cellular differentiation, any dysfunction has significant deleterious consequences to their tissue of residence. It is therefore thought that loss of adult SCs or their functional decline over time results in the tissue and organ dysfunction that is observed in old age. This view is supported by the numerous aging phenotypes described in SCs of the intestine, brain, muscle, skin, germline and hematopoietic system, among others [8]. Although the aging phenotypes of SCs can vary depending on the tissue context, many SCs display conserved characteristics including: support by a niche, unique metabolic requirements, telomerase expression (to prevent replicative exhaustion), cycling between quiescent and activated states, and asymmetric distribution of macromolecules during differentiation [9]. Importantly, these common SC characteristics are known to be perturbed in normal aging by the processes further described in this chapter (e.g. senescence, DNA damage, epigenetics). The role and biology of stem cells in aging has been exhaustively reviewed elsewhere [8, 10–13].

### *3.2.1 Identifying Stem Cells*

The SC definition requires the experimental demonstration that the candidate cell is undifferentiated, able to self-renew (i.e. to divide-indefinitely) and capable to give rise to a tissue's specialized cell types [4, 14]. Experimental approaches such as flow cytometry analysis [15], *in vitro* clonogenicity and differentiation assays [16], immunohistochemistry [17], transplantation studies [18, 19] and genetic lineage tracing [20] have been employed to identify SC populations in most organ and tissue systems. The variability of the SC phenotype has led to the development of specific approaches depending on the tissue in question. Both functional and transcriptomic profiling studies have yielded a plethora of biomarkers used for the isolation and identification of various adult SC populations *in vivo* [17]. It should be noted however, that SC biomarkers should not be over-relied upon, especially in aging studies. Many SC biomarkers are commonly identified in young or developing animals and the utility of these markers in aged organisms should not be assumed to be unequivocal. For example, in the human hippocampus the expression of some neural stem cell/progenitor markers is not affected with age, however a significant decrease directly was found in the expression of the proliferation marker Ki67 and other neuronal markers [21]. This suggests that the simple presence of SC markers is uninformative with respect to their function and that conclusions should not be directly drawn from studies on younger SC compartments.

Following the identification of a potential SC population, they can be transplanted to a host animal to test their ability to self-renew and differentiate [18, 19, 22]. Coupling transplantation with genetic cell labeling allows the tracing of SC fate over time and provides the ability to ascertain a SC's potential for differentiation and self-renewal. Genetic cell ablation studies in which the putative SC is selectively killed can also be conducted to demonstrate tissue loss of homeostatic/regenerative potential [23, 24]. Following this, transplantation experiments can be performed where the candidate SC can be returned to the depleted tissue to test for functional recovery. Experiments involving Cre-Lox genetic lineage tracing [20] and/or cell ablation [25] studies must be careful to investigate whether the promoter selected to drive gene expression is specific to the SC compartment. This prevents potentially labelling/killing differentiated cells and obtaining confounding results. Specificity can be demonstrated by examining the co-localization of reporter expression with SC and differentiation markers by immunohistochemistry soon after activation of the reporter in the putative SC compartment.

### *3.2.2 Technical Considerations for Studying Stem Cell Aging*

Aging can impact SC populations in several ways, including a reduction or increase in SC numbers, reduced proliferative capacity and skewed, or absent, differentiation capacity [8, 12, 13]. These consequences are not mutually exclusive and vary depending on the tissue. SC function is impacted upon by their present intrinsic properties as well as the extrinsic effects of their niche and circulating factors [14]. As these intrinsic and extrinsic factors are affected with age, they are important variables to consider and disassociate when investigating SC aging, especially in the context of cultured SCs in which many extrinsic factors will be unlike those *in vivo*. Comparisons of the impact of aging on adult SCs from different tissues has revealed varying characteristics which are both cell-intrinsic and extrinsic [8].

*In vitro* analysis for speculative SCs is most commonly performed through clonogenicity and differentiation assays. Clonogenicity assays assess the ability of a population of cells to form colonies during *in vitro* culture [16]. However, this technique has several limitations when used in isolation. For example, non-SC populations in the tissue of interest may have sufficient proliferative potential to form colonies or the culture media may not have the necessary composition to promote SC growth. The latter point is also pertinent for *in vitro* differentiation experiments in which SCs are induced to differentiate into a more specialized cell type along a defined lineage. Media composition not mimicking the extrinsic SC niche cues may prevent effective differentiation, or conversely, a cultured cell may be induced to differentiate in an artificial manner not resembling an *in vivo* scenario [26]. In regard to aging studies, providing the same extrinsic factors to young and old SCs may be informative in understanding the intrinsic or extrinsic nature of a given aging phenotype. However, the failure to adequately reproduce the systemic and niche derived factors can result in misleading interpretations. For example, Neural SCs (NSCs) decrease in number during aging, resulting in reduced neurogenesis [27, 28]. Nevertheless, functional *in vitro* analysis did not show significant differences between young and aged NSCs, suggesting that a cell-extrinsic mechanism is involved in their regulation [29].

This notion is supported by experiments conducted through heterochronic parabiosis, an experimental surgical procedure in which two age-mismatched animals have their circulatory systems surgically attached. Using this approach, it was shown that proliferative decay of aged NSCs can be ameliorated by a younger animal´s systemic factors [30–32]. Additionally, heterochronic parabiosis experiments provided evidence that systemic factors can also regulate tissue aging in muscle and liver [33, 34]. A potential improvement to this technique would be to couple it with genetic cell labeling to evaluate blood chimerism or the contribution of SC populations from the experimental animals to tissue regeneration.

Transplantation studies have been a powerful tool in SC biology. They allow the *in vivo* assessment of self-renewal and differentiation of immunophenotypically-defined cell populations through transplantation into their target tissue [18, 19, 22]. This technique has significant advantages over *in vitro* assays due the normal physiological support that an *in vivo* system offers. Aging research has taken advantage of this system by performing transplantation experiments between animals differing in chronological age (known as heterochronic transplantation) [18, 35]. This approach has revealed the importance of intrinsic factors in regulating the function of hematopoietic stem cells during aging [18, 22, 36–41], whereas cell-extrinsic mechanisms predominantly seem to affect the function of satellite SCs in aging skeletal muscle [34, 35, 42, 43].

## 3.3 Genomic Instability, DNA-damage and Telomere Attrition

The occurrence of somatic mutations is common throughout a cell´s lifetime. They can be the result of endogenous events such as DNA replication errors or oxidative stress, or from extrinsic physical, chemical or biological insults [44]. If left unchecked, these errors can lead to gene-specific misregulation, or trigger the activation of cell cycle arrest or cell death pathways. The impact of these events at a cellular level are thought to contribute to the aging phenotype by limiting cellular fitness, depleting stem cells pools and thereby limiting regenerative capacity. This ultimately leads to organ function impairment, predisposing to numerous age-associated pathologies [45]. While the types of lesions arising from these genotoxic insults are highly diverse, complex mechanisms exist to safely repair most of them [44]. Although the accumulation of DNA damage is a hallmark of aging [1], it remains unclear whether it is the result of an increased incidence of genotoxic insults, impaired DNA repair capacity or more likely, a combination of both [46].

DNA damage can be assessed using the comet tail assay [47] or by immunostaining against proteins that accumulate upon DNA damage such as 53BP1 or γH2AX [48]. While global changes in the expression of these proteins can be detected in whole-cell lysates by Western Blot (WB), the visualization and quantification of discrete foci can be achieved by immunofluorescence and high-resolution microscopy [49, 50]. In the case of telomere length, a number of approaches are available including Southern Blotting (SB) of telomere restricted fragments, fluorescent *in situ* hybridization (FISH), qPCR and co-staining of DNA-damage markers with telomere specific proteins (e.g. TRF1 or TRF2) [51].

### 3.3.1 Human Disorders of Premature Aging and DNA-damage Repair

Research exploring many of the premature aging, and DNA-damage repair syndromes in humans has furthered the understanding of the consequences of accumulating DNA damage with age. While the direct contribution of the specific pathognomonic defects to physiological aging remains unclear, collectively they serve as both evidence of a causative role of DNA damage in aging, as well as providing targets for the generation of *in vivo* models to study the aging process (Table 3.1).

### 3.3.2 Animal Models for the Study of Premature Aging Driven by DNA-damage

In terms of modelling human aging *in vivo*, mice are the most commonly used species as they share a similar aging phenotype. They display however, disproportionately long telomeres in comparison to humans, which may limit their relevance in the context of telomere attrition-driven DNA damage. Arguably, simply allowing mice to grow old naturally derives the most physiologically relevant model, but this is time consuming, costly and susceptible to heterogeneity. There are therefore a host of murine models in which this process is accelerated (Table 3.2). Some of these are closely based on the genetics of human premature aging syndromes, such as Hutchinson-Gilford Progeria Syndrome (HGPS) and Werner's Syndrome (WS), which recapitulate many aspects of the human disease and display several specific features of aging. Others are more global models, such as Telomerase Reverse Transcriptase (TERT) deficient mice [59, 60], that generate critically short telomeres after serial generations, or *Bub1b*$^{H/H}$ mice [60], that have impaired mitotic checkpoint function and develop a rapid, global aging phenotype. The multifactorial nature of aging is exemplified by the crossing of telomerase-deficient and WS mice, where the combination of telomere attrition and impaired DNA repair results in a more rapid progeria phenotype than either model alone [62], suggesting that interplay between multiple pathways is responsible for DNA damage-driven aging.

Genomic instability, DNA damage and telomere attrition are core features of aging. While the study of human syndromes of DNA damage and their related animal models has provided useful insights into the phenotypic consequences of these

**Table 3.1** Human premature aging disorders associated with DNA damage

| Name | Predominant clinical features | Genetic lesion | Similarities / Discrepancies with physiological aging (PA) | Reference |
|---|---|---|---|---|
| **Laminopathies** | | | | |
| Progeria (Hutchinson-Gilford Progeria Syndrome (HGPS) | Growth impairment, cardiovascular disease, skeletal dysplasia, lipodystrophy, alopecia, skin and nail defects, joint contractures, premature death (2nd–3rd decade). | Sporadic autosomal dominant point mutations in the LMNA gene, resulting in activation of a cryptic splice site and an in-frame 50AA deletion. | HGPS is a prototypic premature aging syndrome. | [52, 53] |
| **RecQ disorders** | | | | |
| Werner's Syndrome (WS) | Short stature (absent adolescent growth spurt), skin atrophy, bird-like faces, lipodystrophy, hair greying, cataracts, Achilles tendon ulceration, type 2 diabetes, cardiovascular disease, osteoporosis, hypogonadism, malignancy, premature death (5th to 6th decade). | Loss of function (LOF) mutation in WRN gene | Generally considered a premature aging syndrome. Minor disparities include: 1) Cataracts seen in WS are typically posterior sub-capsular cf. nuclear with PA. 2) The malignancies seen with WS are not those typical of PA. 3) The increased CVD seen is WS is not associated with hypertension as it is in PA. 4) Osteoporosis is more common in distal limbs in WS rather than the vertebral column as in PA. | [54] |
| **Telomeropathies** | | | | |
| Dyskeratosis Congenita | Core features (≥80%): leukoplakia, nail dystrophy, hyperpigmentaion, bone marrow failure, premature death. | Numerous mutations: Mutations in DKC1 gene, a telomerase component gene (TERC, TERT, NOP10, NHP2, or TCAB1) or a shelterin component (TINF2). | DKC encompasses a spectrum of disorders with the common feature of impaired telomere maintenance. As a result, this group is clinically heterogeneous, and while many features of DKC mimic PA, the predominant feature, and most frequent cause of death is bone marrow failure. | [55, 56] |

(continued)

**Table 3.1** (continued)

| Name | Predominant clinical features | Genetic lesion | Similarities / Discrepancies with physiological aging (PA) | Reference |
|---|---|---|---|---|
| **Other DNA damage syndromes** | | | | |
| Cockayne syndrome (CS) | Cachectic dwarfism, severe neurological manifestations (microcephaly, cognitive deficit), cataracts, sensorineural deafness, pigmentary retinopathy, photosensitivity, joint contractures, accelerated hypertension, aortic root dilatation and cardiomyopathy, premature death (2nd decade) | Two main groups: CSA due to LOF mutations in ERCC8; CSB due to LOF mutations on ERCC6. | Hearing loss and deafness in CS are akin to PA, as is hypertension with its associated end organ damage. Although mitochondrial pathology is a feature of PA, it appears significantly accelerated in CS. No associated increase in malignancy in CS in contrast to other nucleotide excision repair defects and PA. | [57] |
| Fanconi anemia (FA) | Myelodysplastic syndrome, bone marrow failure, acute myeloid leukemia, osteoporosis, sarcopenia, immune deficiency, endocrine dysfunction, increased susceptibility to malignancy. | 19 implicated genes, principally DNA repair genes. | FA is a collective term for a clinically heterogeneous group. The predominant feature of FA is BMF, MDS and AML. Typically, MDS and AML are PA associated diseases with median onset of 50 and 70 years respectively in PA, but these occur prematurely (10 and 30 years) in FA. FA therefore recapitulates bone marrow aging in compartmental fashion rather than global aging. | [58] |

Related syndromes/disorders that are not completely considered to reflect premature aging or DNA damage have not been included. HGPS, Hutchinson-Gilford Progeria Syndrome; PA, physiological aging; WS, Werner's Syndrome; CVD, cardiovascular disease; DKC, Dyskeratosis Congenita; CS, Cockayne syndrome; LOF, Loss of function; FA, Fanconi anemia; BMF, bone marrow failure; MDS, myelodysplasia; AML, acute myeloid leukemia

**Table 3.2** *In vivo* models of DNA-damage driven premature aging

| Disease /Target | Genotype /Lesion | Phenotype | Reference |
|---|---|---|---|
| **Disease Specific Models** | | | |
| Laminopathy models | | | |
| HGPS | *Lmna*$^{HG}$ knock-in mice. Accumulation of farnesylated, uncleaved progerin. | Heterozygote (*Lmna*$^{HG/+}$) shows: phenotype onset by 6–8 weeks, premature death (50% by 27 weeks), lipodystrophy, kyphosis, osteoporosis, hair loss, low birth rate. Homozygote (*Lmna*$^{HG/HG}$) shows: low birth rate, severe skeletal abnormalities, death by 3–4 weeks. | [63, 64] |
| HGPS | Lmna c.1827C>T;p. Gly609Gly. Point mutation resulting in splice variant akin to HGPS. Progerin accumulation. | Heterozygote (*Lmna*$^{G609G/+}$) shows: phenotype onset by 8 months, death by 7–9 months. Homozygote (*Lmna*$^{G609G/G609G}$) shows: phenotype onset from 3 weeks, death by 3–4 months. General features: impaired growth, lordokyphosis, osteopenia, lipodystrophy, cardiovascular dysfunction, vascular calcification, endocrine dysfunction, accumulation of senescent cells, accumulation of cellular DNA damage. | [65, 66] |
| DNA repair syndromes | | | |
| WS | *Wrn*$^{\Delta hel/\Delta hel}$. Homozygous helicase domain deletion. (Note that *Wrn*$^{null/null}$ do not exhibit a phenotype). | Mild premature aging phenotype, higher DNA mutation rate, higher reactive oxygen species levels, hallmarks of a metabolic syndrome (visceral obesity, hypertriglyceridemia, type 2 diabetes, increased cardiovascular risk), increased rate of malignancy, limited lifespan (10-15% shorter than controls). | [67–70] |
| WS | *Wrn*$^{\Delta hel/\Delta hel}$ / *Terc*$^{-/-}$. Homozygote helicase deletion crossed with telomerase deficient mice, manifesting in critical short telomeres. | Initial phenotype features from 4 months in *G4-6Terc*$^{-/-}$ / *Wrn*$^{\Delta hel/\Delta hel-}$ and include: hair greying, alopecia, osteoporosis, type 2 diabetes, cataracts, impaired wound healing, impaired glucose tolerance, increased rate of malignancy, limited lifespan (median survival 24 weeks compared to 96 weeks in control). | [62] |
| CS | *Csb*$^{m/m}$. Homozygous truncation of the CBS (ERCC6) gene resulting in CBS deficiency. | Deficient TC-NER photosensitivity, growth impairment, mild neurological dysfunction, deafness, age-dependent blindness, increased susceptibility to skin cancer, normal lifespan. | [71, 72] |

(continued)

**Table 3.2** (continued)

| Disease /Target | Genotype /Lesion | Phenotype | Reference |
|---|---|---|---|
| **Non-disease specific models** | | | |
| Telomerase deficient mice | *mTR*$^{-/-}$. Homozygote deletion of the telomerase RNA component. Telomere attrition manifesting with critically short telomeres by generation 4-6. | Sequential telomere shortening with generations. No phenotype in early generations. Later they show: grey hair, alopecia, skin ulceration and dermal fibrosis, weight loss, increased susceptibility to cancer, impaired wound healing, impaired stem cell repopulation capacity, myeloid skewing, limited lifespan (18 months vs 24 in control). | [59, 60, 73–76] |
| Bub1b (DNA replication/cell cycle checkpoint) | *Bub1b*$^{H/H}$. Hypomorphic Bub1b. Note that *Bub1b*$^{-/-}$ is embryonically lethal, while *Bub1b*$^{-/H}$ die perinatally. | Impaired checkpoint and DNA repair, accumulation of DNA damage. Normal at birth, onset at 3–4 months, progressive aneuploidy, cachectic dwarfism, lordokyphosis, lipodystrophy, cataracts (nuclear), impaired wound healing, infertility, reduced lifespan (median 6 months). | [61] |
| Mitochondrial DNA damage | PolgA$^{mut}$ /PolgA$^{mut}$ knock-in. | Results in accumulation of mitochondrial DNA damages and DNA deletion. Normal at birth, onset 25 weeks, kyphosis, alopecia, weight loss, osteoporosis, anemia, cardiomyopathy, impaired fertility, increased mitochondrial mass and impaired mitochondrial function, reduced lifespan (median 48 weeks). | [77] |

HGPS, Hutchinson-Gilford Progeria Syndrome; CVD, cardiovascular disease; DNA, deoxyribonucleic acid; WS, Werner's Syndrome; CS, Cockayne syndrome; TC-NER, transcription-coupled nucleotide excision repair; NER, nucleotide excision repair; AT, Ataxia-telangiectasia; RNA, Ribonucleic acid

aberrations, the process of aging is complex and multifactorial, meaning that there is not (nor is there ever likely to be) a single robust model.

## 3.4 Epigenetics of Aging

Epigenetic mechanisms can control gene expression in a heritable manner without altering the underlying DNA sequence. Histone modifications, chromatin reorganization, DNA methylation and other epigenetic mechanisms are capable of

**Table 3.3** Epigenetic alterations in aging, with comments on experimental data sources and model relevance

| Epigenetic alteration | Description / Evidence of relevance | Model system note | Reference |
|---|---|---|---|
| **Histone Loss** | General loss of histones in aging cells across multiple species. Overexpression of histone in yeast extends lifespan. | Observational data from human and murine samples. Functional data from yeast. | [79, 80] |
| **Histone modifications** | **Methylation** Yeast and worms show global decrease in repressive (H3K27me3) and increase in activating methylation marks (H3K4me3), with resulting changes in gene expression in aging. | The changes observed in yeast and worms are often not recapitulated in flies and mammals, raising questions about their validity in those systems. The relationship between histone methylation and aging in flies and mammals is less clear. | [82, 102] |
| | **Acetylation** Increased acetylation is observed with aging. Overexpression of deacetylases extends lifespan. | The global changes in acetylation seen are conserved across species. Mouse models of overexpression of sirtuin show varying degrees of resistance to aging, and depletion of SIRT6 results in a progeria phenotype. | [83, 85, 87, 88, 103–107] |
| **Chromatin re-organization** | Global heterochromatin loss, with focal reorganization and SAHF formation in senescent cells. | Human aging and progeria models show global heterochromatin loss with focal reorganization. Senescent cells undergo chromatin reorganization and SAHF formations, directly contributing to the phenotype. Modulating HP1 in *Drosophila* alters lifespan. | [92, 94] |
| **DNA Methylation** | Global hypomethylation associated with aging, but with regions of focal hypermethylation. No evidence that manipulating DNA methylation alters aging or lifespan. | Lower complexity organisms (yeast, worms and flies) have no, or limited, DNA methylation. Mammalian systems are therefore better placed to study DNA methylation. All findings are observational. | [95, 96] |

DNA, deoxyribonucleic acid; SAHF, senescence-associated heterochromatin foci; HP1, heterochromatin protein 1

dynamically altering gene expression in virtually every cell type and tissue. A host of epigenetic changes occur with aging (Table 3.3) [78], and these have attracted particular attention because of their reversible nature. This makes them potential therapeutic targets for limiting the effects of aging and extend both health- and lifespans.

### 3.4.1 Histone Alterations

Age-associated loss of histone proteins has been described in a number of organisms, including humans [79, 80]. While the mechanisms controlling this global histone loss remain unclear, it has been possible to show that increasing the histone supply extends cell lifespan in yeast, suggesting that the histone dosage is critical [79].

Global changes in histone modifications can be assessed with modification-specific antibodies visualized by immunostaining or WB. Their functional effects however, are best characterized by identifying genes, promoters or other DNA regions with which they interact. This is best achieved with chromatin immunoprecipitation (ChIP) and qPCR (ChIP-PCR) or Next Generation Sequencing (NGS) (ChIP-seq).

The modification of histones by the addition or removal of a methyl or acetyl group directly regulates gene expression. Histone methylation can have either activating or repressive effects on gene expression. In the nematode worm *C. elegans*, researchers found that aging involved a global gain of activating histone methylation marks (H3K4me3) and loss of repressive marks (H3K27me3), and that restoring their normal levels could extend lifespan [81, 82]. However, the global gain of activating marks and loss of repressive ones in aging does not appear to be preserved in other species such as flies or mammals [78].

In contrast to histone methylation, the relationship between histone acetylation and aging is better established. Evidence suggests that acetylation increases with age and inducing hypoacetylation extends life- and healthspan. For example, spermidine is a naturally occurring polyamine that induces histone H3 deacetylation through inhibition of histone acetyltransferases. The levels of spermidine decline with age, and administration of supplemental spermidine extends lifespan in yeast, flies, cultured human immune cells and mice [83]. In addition, the role of sirtuins, which are NAD-dependent protein deacetylases, has also been extensively studies in yeast, worms and mammals. The overexpression of Sir2 was first shown to extend lifespan in yeast [84], then worms [85] and flies [86]. Mammalian models also seem to show a similar trend, as mice overexpressing different sirtuins display phenotypes related to delayed aging as improved health, resistance to DNA- and metabolic damage, diminished age-associated changes in the HSC compartment and even lifespan extension [87–89].

### 3.4.2 Chromatin Reorganization

The physical structure of chromatin has significant impact on gene transcription [90]. Heterochromatin is tightly packed DNA and while there are exceptions, these regions are typically not transcribed. Besides histone-modifying enzymes, a number of other factors are capable of controlling heterochromatin and nucleosome organization. Notably, some of these are specifically altered in aging, including heterochromatin protein 1a (HP1a), polycomb group proteins and the NuRD complex [91]. The 'loss of heterochromatin' model of aging proposes that heterochromatin domains established during embryogenesis are gradually lost with age, leading to transcription of age-associated genes. The role of impaired heterochromatin maintenance aging is supported by data from some of the progeria syndromes and functional studies in flies [92, 93]. However, it is important to note that this model of global loss of heterochromatin is an oversimplification. As an example, one of the most striking features of senescent cells is their chromatin reorganization into senescence-associated heterochromatin foci (SAHF), which can be directly observed with DNA-specific fluorescent dyes such as 4′6-diamino-2-phenylindole. SAHF occur in a coordinated fashion, and directly contribute to the establishment of the senescence phenotype, by regulating specific target genes [94]. The relevance of senescent cells in aging is discussed in Sect. 3.7.

### *3.4.3 DNA Methylation*

In addition to histone modifications, DNA can be directly modified to regulate expression. Methylation of DNA occurs at cytosine-guanine rich regions (known as CpG islands) and is associated with repressed gene expression. While the techniques to study DNA methylation (e.g. bisulfite conversion and sequencing) are relatively straightforward [95], many of the model organisms used to study aging (e.g. yeasts, worms and flies) have little or no DNA methylation [78], significantly hindering functional studies.

The most compelling evidence of an association with aging and DNA methylation comes directly from human studies. With increasing age, mammalian cells undergo global DNA hypomethylation with focal areas of hypermethylation [95], particularly at tumor suppressor genes and polycomb targets [97]. Many of these features are largely recapitulated in mouse and human progeria syndromes [98, 99], and it has been shown that DNA methylation profiles can accurately predict chronological age, age-related pathologies and mortality [100, 101]. This data is however observational, and no experimental data currently exists supporting a direct causative association between DNA methylation and aging.

## 3.5 Deregulated Nutrient Sensing

Nutrient sensing is a well-orchestrated, evolutionary conserved process that is essential for the survival of all living beings [108]. Whether it is a simple prokaryote or a complex eukaryote, all organisms have the ability to not only sense changes in environmental cues, but also to efficiently utilize nutrients present in the environment for the generation of cellular energy and the building blocks of cells [109]. A host of nutrient sensing pathways ensure a fine balance between anabolism when nutrients are abundant, and catabolism, during times of nutrient scarcity. Therefore, it is unsurprising that deregulated nutrient sensing is implicated in a range of pathologies (e.g. obesity, diabetes and cancer) and that is also considered one of the Hallmarks of Aging [1]. Nutrient sensing mechanisms vary greatly, and in animals they can range from insulin and insulin-like growth factor signaling (IIS) to the mechanistic target of rapamycin (mTOR) pathway and the somatotropic axis [108, 109]. The IIS and mTOR pathways will be the primary focus of this section, as are tightly linked and are amongst the most evolutionary conserved pathways that are involved in regulating the aging process.

A variety of dietary, genetic and pharmacological approaches have been utilized to study the role of nutrient sensing in aging across a range of species. Dietary restriction is perhaps the most robust approach to improve healthspan so far, as it has been successful in doing so in all tested organisms including primates [110]. In addition, intermittent fasting and the recently demonstrated fasting-mimicking diet (FMD) have also emerged as potential avenues to improve various aspects of health span, including potentially reversing diabetes and ameliorating dementia and cancer [111–113]. Importantly, intermittent fasting also improved several markers and risk factors of aging and age-related disease in humans [114]. However, several caveats remain for interpreting the data in order to understand how these approaches can be applied efficiently in a clinical setting. Many outcomes from these fasting regimens and FMDs will depend on the duration and severity of the fasting regimen, the type of diet used to mimic fasting and the time of onset of the dietary regimen. This is crucial since severe caloric restriction could lead to malnutrition and result in deleterious effects such as immunosuppression and reduced fecundity. Another major hurdle is the practicality of adhering to strict fasting regimes for prolonged periods. Thus, there is still a need for pursuing alternative avenues. To this end, genetic and pharmacological approaches targeting key regulators of metabolic processes such as mTOR or its downstream target, ribosomal protein S6 kinase 1, may pave the way for future anti-aging therapeutics [115, 116].

Rapamycin is a naturally occurring inhibitor of mTOR and rapamycin treatment is perhaps the most robust chemical intervention able to prolong lifespan in various organisms [117]. Similar to dietary regimens, *in vivo* chemical inhibition experiments require careful consideration of dosage, length of treatment and the onset of the chemical intervention, as these will affect the robustness of the findings and their relevance for human translation. Nevertheless, rapamycin treatment was shown to extend lifespan even if given intermittently in middle-aged mice, suggesting that adult-onset treatment is sufficient [115, 118]. Important factors to consider are feeding time of the day and fed status at the moment of sacrifice. In this case, over-night fasting provides a common circumvention that ensures all animals will be at a baseline status for tissue analysis. The genetic background and housing hygiene can also have profound effects on the observed phenotype in long-term metabolic and aging experiments where the immune system can have a major influence. It should also be unsurprising that manipulating many of these metabolic pathways can result in sexually dimorphic phenotypes, thus it is imperative for any aging study to study both sexes. For example, rapamycin increased lifespan in females at lower doses than males, possibly due to sexual dimorphisms in drug metabolism [119]. Therefore, similar chemical interventions should involve appropriate pharmacodynamic and pharmacokinetic evaluations at systemic and target tissue levels across sexes. To further highlight this, high doses of intermittent rapamycin treatment to increased male lifespan, whereas it failed to do so in females and also shifted cancer incidence towards a highly aggressive hematopoietic malignancy [118].

In addition to chemical interventions, genetic manipulation of several conserved nutrient sensing pathways has been shown to increase lifespan in different model species such as yeast, worms, flies and mice [116, 120–123]. Nevertheless, some limitations to these studies must be noted from a methodological perspective. In the case of genetically modified mice, many of the reported genetic strategies (e.g. gene deletions) are already present during embryonic development, which means that some of the observed phenotypes may be confounded by effects on both embryonic and early-life development, especially as many of these genes are essential for general metabolism and nutrient sensing throughout life. Moreover, these effects might render the experimental strategy unviable. As an example, mice with complete knockout of mTOR display embryonic lethality. Therefore, future approaches should investigate the effects of adult-onset deletion/manipulation of key genes on mammalian lifespan, as this would be the basis for future research exploring therapeutic intervention in humans. Another useful strategy for circumventing embryonic/early life effects is to generate hypomorphic alleles where gene expression is significantly diminished but not absent. As an example, mTOR hypomorphic mice are viable and show a significant extension in lifespan in both genders [124].

## 3.6 Age-related Chronic Inflammation

Chronic, low-level inflammation in the absence of infection (known as "inflammaging") is tightly related to many age-related pathologies, including frailty syndrome, diabetes and cancer [1, 124, 125]. In contrast to an acute inflammatory response that is beneficial and promotes tissue repair, this persistent, smoldering inflammation associated with aging is deleterious and can cause tissue deterioration. Inflammaging is characterized by the elevated secretion of several proinflammatory factors (e.g. IL1, IL6, IL8, TNF-$\alpha$, C-reactive protein and reactive oxygen species) in all tissues. A diverse range of stimuli is thought to contribute to age-related tissue and systemic inflammation such as accumulation of genomic damage, the senescence-associated secretory phenotype (SASP), a defective autophagic response as well as an exhausted immune system that can no longer efficiently clear pathogens and dysfunctional host cells [126].

There is a considerable amount of clinical data implicating chronic inflammation in aging, including that it can predict changes in body composition, metabolic balance, energetic consumption and immune response capacity, besides being a highly significant risk factor for mortality and disease in older adults [126, 127]. Nevertheless, there is a scarcity of experimental evidence supporting inflammaging as a driver of aging and its related diseases. There are however, findings providing important causal data. A notable example lies in the NLRP3 inflammasome, a protein complex activated by a diverse range of age-dependent "danger signals", including lipotoxic free acids, extracellular ATP and reactive oxygen species. Importantly, *Nlrp3*$^{-/-}$ mice show increased healthspan with improved glucose homeostasis, bone density, muscle endurance as well as protection against immunosenescence (loss of naïve T-cells and B-cells) [128, 129]. In the future, it will be particularly interesting to study the role of inflammatory factors through heterochronic parabiosis experiments, which have previously illustrated that secreted factors present in young blood can rejuvenate cardiac, muscular and cognitive functions in old mice [32, 34, 130–132] (see Sect. 3.2, for further discussion on stem cells).

In any case, delineating the sources and causes of age-related inflammation remain major issues yet to be addressed. In this sense, cellular senescence has recently caught the field's attention as a potential culprit, or at least contributor, to age-related inflammation [126]. As further described in Sect. 3.7, senescent cells secrete a vast array of proinflammatory factors and eliminating senescent cells in aged mice through semi-genetic or pharmacological approaches leads to the reduction of proinflammatory factors Il6, Il1a and Tnf [133, 134]. Moreover, the NLPR3 inflammasome also regulates oncogene-induced senescence and its SASP, therefore some of the benefits of the global NLRP3 deletion may be attributed to preventing senescence in aging [135].

Age-related inflammation can be assessed by a variety of methods, but it is important to combine several approaches to obtain a broader view at tissue and organismal levels. For example, a common method to evaluate inflammation in aged mammals is to measure pro-inflammatory cytokine levels and absolute number of immune cells in the peripheral blood. However, these levels display significant variability among individuals as expected from naturally aged mice, thus a considerable sample size is needed to detect statistically significant differences. Therefore, the inflammatory response of the tissue of interest should also be assessed. This can be achieved by measuring local gene expression levels of various chemokines, cytokines and surrogate markers of immune cells by quantitative polymerase chain reaction (qPCR), WB, enzyme-linked immunosorbent assay or liquid chromatography coupled with mass spectrometry. Additionally, histological assessment of immune cell infiltration can also be conducted by immunostaining against common markers such as CD3 (T cells), B220 (B cells), F4/80 (macrophages), Ly6G (neutrophils) and MHC II [136]. This can be combined with fluorescence-activated cell sorting (FACS) to evaluate various immune cell types in more detail. Although observing cytokine and chemokine expression at the tissue level has historically proven difficult to achieve, novel variants of the RNA *in situ* hybridization technique have provided important solutions to this caveat. One example is the RNAscope® platform, which can be combined with cell-specific immunostaining in order to identify which cell type is responsible for secreting proinflammatory factors of interest in aged tissues [137].

Overall, chronic inflammation is rapidly being recognized as an important driver of aging and age-related pathology, though there is still much to learn about the nature and origin of this inflammatory response. Novel animal models aimed at manipulating crucial inflammatory pathways will surely provide essential insight into the relationship between inflammation and health span, as well as providing evidence supporting therapeutic targeting of this deleterious inflammatory response.

## 3.7 Cellular Senescence in Aging

Senescence is a cellular state characterized by an irreversible cell cycle arrest and altered gene expression that has been demonstrated to act as a potent tumor-suppressive mechanism [138, 139]. This phenomenon is induced by various damaging cellular stresses including telomere shortening, a persistently activated DNA-damage response [140] and oncogene activation [141]. Senescent cells are known to induce pleiotropic effects on neighboring cells through an extensive secretome of pro-inflammatory cytokines, chemokines, extracellular matrix proteases and growth factors, which is collectively known as the SASP [142]. While canonically considered an anti-tumoral mechanism preventing the division of damaged cells, recent research has highlighted the involvement of senescent cells in numerous pathological processes, mainly aging and age-associated diseases [143]. Notably, senescent cells are observed to over accumulate in most aging tissues. This has been well characterized in the lung, liver, skin and spleen of aging mice, primates and humans [144–146].

The role of cellular senescence during the normal aging process is currently debated with two main hypotheses being considered. The first postulates that the accumulation of senescent cells and their SASP result in tissue dysfunction leading to an aging phenotype. The second view is that senescence may reduce the regenerative capacity of adult stem cells, which are necessary for tissue homeostatic balance. These hypotheses are not necessarily mutually exclusive and may act concomitantly during normal aging. Evidence supporting senescence having a causative role in aging is derived from observations that p16/INK4A-deficient mice, in which the senescence program is abrogated, have increased lifespans and reduced incidence of age-associated disease. Notably, this effect is also observed in progeroid or normal mice when senescent cells are ablated genetically or with chemical compounds that specifically target senescent cells (i.e. senolytics) [147–151]. However, further work is required to determine the mechanisms by which senescent cells accumulate during aging and their specific effects on the aging tissue microenvironment. Until recently, this area of study was confounded by the lack of available biomarkers and tools to study these processes *in vivo*.

### 3.7.1 Cellular Senescence: Lessons From In vitro Research

Over the past 50 years, *in vitro* research of cellular senescence has established essential techniques and principles for the study this phenomenon. As senescent cells cannot be grown in culture, an initial challenge was to discover means to induce senescence *in vitro*. A frequently employed technique is the serial passage of primary cells until they reach replicative exhaustion, which triggers telomere attrition, DNA damage and senescence [152]. Other strategies aimed at inducing either widespread or telomere-specific DNA damage, such as ionizing irradiation [140], DNA-damaging drugs [153] or depletion of the shelterin complex [154] are also robust inducers of senescence.

The detection of senescent cells relies on markers that reflect several altered cellular processes. For example, their inability to proliferate is mainly due to elevated expression of cyclin-dependent kinase inhibitors (CDKi) (e.g. p21/Cip1 and p16/INK4A). Other features include cell volume enlargement and expansion of the lysosomal compartment, which is reflected by the increased activity of Senescence-Associated β-Galactosidase (SA-β-Gal) [146]. The latter is the most widely used marker of senescence, although it is not infallible. Senescent cells are also observed to up regulate pro-survival and anti-apoptotic factors [155–159], as well as markers of DNA damage [160]. Finally, senescent cells activate the SASP as a downstream consequence of NF-κB and p38-MAPK signaling [161–163]. The complex nature of this phenotype leads then to a crucial methodological consideration: currently there is no single method able to unequivocally identify senescent cells [164].

### 3.7.2 Common Methods for the In vivo Identification of Senescent Cells

During the normal aging process, senescent cells are found to accumulate in tissues as evidenced by the increased detection of cells with high SA-β-Gal activity and elevated expression of p16/INK4A [146, 165–168]. To date, the best practice for the identification of senescent cells *in vivo* is combining the use of the aforementioned markers plus demonstrating absence of proliferation. However, results obtained from any single marker should not be deemed definitive, as *bona fide* senescent cells can sometimes lack even some of the most robust markers [169].

Initially, bulk cell populations or tissues can be analyzed for SASP factors, elevated SA-β-Gal activity and CDKi expression by WB or qPCR. This approach has been widely used for demonstrating the presence of senescent cell populations *in vivo* [146]. However, such strategies have significant limitations and results should be interpreted cautiously. For example, many non-senescent immune cells can display a pro-inflammatory secretome resembling the SASP. The composition of the SASP can also be highly heterogeneous depending on the senescence-inducing stimuli or the "maturity" of the senescent cell, as is the case of some progeroid mouse strains which show elevated levels of senescence in their adipose tissues but with different SASP composition [147, 170]. Also, it is currently unknown if the length of time following the onset of cellular senescence has any qualitative of quantitative effect on the SASP. Therefore, further work is required to characterize the causes and effects of SASP heterogeneity in different cellular and temporal contexts.

The identification of senescent cell populations through analysis of CDKi expression also has several limitations. The most widely used markers, p21/Cip1 and p16/INK4A, are also expressed by many quiescent cell populations [9, 171] or aged immune cells [172]. In the case of p16/INK4A, which is considered a more robust senescence marker, the lack of reliable antibodies has complicated its detection both *in vivo* and *in vitro*. High levels of SA-β-Gal activity are also found in maturing macrophages that can display a pro-inflammatory, SASP-like expression signature, possibly to misleading results when analyzing tissues for the presence of cellular senescence [146]. Following the identification of potential senescent tissues, the senescent cell type should be ascertained through histological approaches such as SA-β-Gal staining multiplexed with immunostaining for specific cell-type markers, as well as lack of proliferation.

### *3.7.3 Novel Tools and Models for the Study of Cellular Senescence*

The *in vivo* identification and study of senescent cells has benefited considerably from the recent development of genetically engineered mice in which a traceable marker is expressed in cells that activate the senescence program. These "senescence-reporting mice" include p16-3MR [173], INK-ATTAC [147] and p16-LUC [165]. It should be taken into account that all of these models are based on the expression of *p16/INK4A* as a senescence biomarker, as this gene can be expressed in non-senescent cells. Furthermore, each model has intrinsic advantages and disadvantages that should be weighted according to the experimental context. p16-LUC mice contain a luciferase knocked-in downstream of the start codon of one endogenous p16/INK4A allele [165]. These mice allow for whole-body luciferase imaging, which permits the non-invasive global identification of senescent cells during aging. However, the luciferase knock-in results in disruption of one copy of p16/INK4A, which is a confounding variable that must be considered alongside appropriate controls (e.g. comparing with p16+/- animals).

The other two models, p16-3MR and INK-ATTAC, not only allow the detection of senescent cells but also permit specifically ablating them by administering certain drugs. However, the deleterious effects of continuous drug administration must be taken into account for their use in longevity studies. As an advantage, p16-3MR and INK-ATTAC do not result in the disruption of one of the endogenous p16/INK4A alleles, while p16-3MR can also be used to non-invasively identify bulk cellular senescence in tissues due to the expression of luciferase. Another consideration is that since both p16-3MR and INK-ATTAC use different promoter elements, it may be possible that their expression differs depending on cellular context and/or senescence-inducing stimuli. The mechanism of cell ablation in both models also differs, meaning that each may be more efficient at inducing apoptosis depending on the context. As both p16-3MR and INK-ATTAC harbor fluorescent reporters, FACS isolation of senescent cell populations can be performed, which can be further investigated by qPCR and WB for SASP expression. This can permit a more refined comparison of the senescent state across varying age and tissue. Furthermore, single-cell RNA sequencing can also be performed, allowing for data to be obtained on the transcriptomic heterogeneity of individual senescent cells.

New non-genetic tools for studying cellular senescence have also been generated. For example, mesoporous silica nanoparticles capped with galacto-oligosaccharides (GOS) have been used *in vitro* to label senescent cells [174]. These nanoparticles take advantage of the increased β-Galactosidase activity of senescent cells to remove the GOS cap, specifically releasing their cargo within them. The nanoparticles can be filled with dyes such as rhodamine to allow identification of the senescent cells or with genotoxic drugs to ablate them. This tool is potentially advantageous, as it removes the need for breeding of senescence-reporters onto experimental genetic backgrounds and it possesses versatility as the nanoparticles can be filled with a variety of molecules for probing the location and function of senescent cell populations. However, further research is required to ascertain their utility *in vivo,* while the effects of off-target cargo release should also be characterized (for example in macrophages), as they could potentially impact health and lifespan in aging studies.

## 3.8 Conclusions

Biomedical research has been instrumental in developing our understanding of the molecular and cellular hallmarks that characterize aging, which are conserved among humans and various other species. During the last decade, both *in vivo* and *in vitro* experimental approaches have been applied in these model species, in order to characterize the role of these hallmarks, as well as their underlying mechanisms. However, many challenges remain for basic aging research from a methodological standpoint. An obvious caveat is the fact that lower life forms lack the complexity of higher mammalian systems. Therefore, some of these models lack one or more features that are considered of importance for aging in humans, such as the absence of a complete immune system. Moreover, many of these studies have been conducted *in vitro*. These systems can certainly be useful in specific contexts, such as cell-type specific models for high throughput screening for the discovery of new target genes and pathways, or the identification of novel bioactive drugs. However, in practical terms, the interpretation of findings produced in such systems should be restricted to the processes that can be addressed by the experimental setup. Given that aging is a multifaceted, multi-system disorder, *in vivo* systems remain the most pertinent models for its study from an organismal perspective. Therefore, an important next step will be the *in vivo* demonstration of results that have been mainly produced in culture. Still, living animal models also have their limitations, one being the highly time-consuming nature of chronological aging experiments in mammalian species, our closest relatives. This difficulty has largely prevented the independent routine validation of results by different groups in diverse locations, an aspect of great importance given the inherent variability of aging studies. However, the greatest challenge will be developing a unified understanding of the role of the Hallmarks of Aging in longevity and age-related disease, as most of these discoveries have been produced in particular mutant animals or experimental models. Finally, it must be noted that translational applications are still in their infancy in this discipline. Nevertheless, the number of clinical trials as well as biotech companies involved in developing anti-aging drugs and therapeutics has grown considerably [175], remarking the importance of current and future biomedical research in aging (see Chapter 17 for a complete description of Future Research in Aging).

**Acknowledgements** We are very grateful to Prof. Juan Pedro Martinez-Barbera for critical reading and comments on the manuscript.

## References

1. López-Otín C, Blasco MA, Partridge L, Serrano M, Kroemer G (2013) The hallmarks of aging. Cell 153(6):1194–1217. https://doi.org/10.1016/j.cell.2013.05.039
2. Taylor RC, Dillin A (2011) Aging as an event of proteostasis collapse. Cold Spring Harb Perspect Biol 3(5):a004000. https://doi.org/10.1101/cshperspect.a004440
3. Baumann K (2016) Ageing: the yin and yang of mitochondrial dysfunction. Nat Rev Mol Cell Biol 17(6):331. https://doi.org/10.1038/nrm.2016.71
4. Hsu Y-C, Li L, Fuchs E (2014) Emerging interactions between skin stem cells and their niches. Nat Med 20(8):847–856. https://doi.org/10.1038/nm.3643
5. Patel DM, Shah J, Srivastava AS (2013) Therapeutic potential of mesenchymal stem cells in regenerative medicine. Stem Cells Int 2013:1–15. https://doi.org/10.1155/2013/496218
6. van Es JH, Sato T, van de Wetering M, Lyubimova A, Nee ANY, Gregorieff A et al (2012) Dll1+ secretory progenitor cells revert to stem cells upon crypt damage. Nat Cell Biol 14(10):1099–1104. https://doi.org/10.1038/ncb2581
7. Gómez-Gaviro MV, Lovell-Badge R, Fernández-Avilés F, Lara-Pezzi E (2012) The vascular stem cell niche. J Cardiovasc Transl Res 5(5):618–630. https://doi.org/10.1007/s12265-012-9371-x
8. Schultz MB, Sinclair DA (2016) When stem cells grow old: phenotypes and mechanisms of stem cell aging. Development 43(1):3–14. https://doi.org/10.1242/dev.130633
9. Cheung TH, Rando TACTH (2013) Molecular regulation of stem cell quiescence. Nat Rev Mol Cell Biol 14(6):329–340. https://doi.org/10.1038/nrm3591
10. Sharpless NE, DePinho RA (2007) How stem cells age and why this makes us grow old. Nat Rev Mol Cell Biol 8(9):703–713

11. Childs BG, Durik M, Baker DJ, van Deursen JM (2015) Cellular senescence in aging and age-related disease: from mechanisms to therapy. Nat Med 21(12):1424–1435. https://doi.org/10.1038/nm.4000
12. Oh J, Lee YD, Wagers AJ (2014) Stem cell aging: mechanisms, regulators and therapeutic opportunities. Nat Med 20(8):870–880. https://doi.org/10.1038/nm.3651
13. Ahmed ASI, Sheng MH, Wasnik S, Baylink DJ, Lau K-HW (2017) Effect of aging on stem cells. World J Exp Med 7(1):1–10. https://doi.org/10.5493/wjem.v7.i1.1
14. Morrison SJ, Spradling AC (2008) Stem cells and niches: mechanisms that promote stem cell maintenance throughout life. Cell 132(4):598–611. https://doi.org/10.1016/j.cell.2008.01.038
15. Donnenberg VS, Ulrich H, Tárnok A (2013) Cytometry in stem cell research and therapy. Cytometry A 83(1):1–4. https://doi.org/10.1002/cyto.a.22243
16. Sarma NJ, Takeda A, Yaseen NR (2010) Colony Forming Cell (CFC) assay for human hematopoietic cells. J Vis Exp 18(46):pii2196. https://doi.org/10.3791/2195
17. Pavlović M, Radotić K (2017) Stemness and stem cell markers. In: Animal and plants stem cells. Springer International Publishing, Cham, pp 27–32
18. Ogden DA, Mickliem HS (1976) The fate of serially transplanted bone marrow cell populations from young and old donors. Transplantation 22(3):287–293
19. Beauchamp JR, Morgan JE, Pagel CN, Partridge TA (1999) Dynamics of myoblast transplantation reveal a discrete minority of precursors with stem cell-like properties as the myogenic source. J Cell Biol 144(6):1113–1122
20. Hsu Y-C (2015) Theory and practice of lineage tracing. Stem Cells 33(11):3197–3204. https://doi.org/10.1002/stem.2123
21. Mathews KJ, Allen KM, Boerrigter D, Ball H, Shannon Weickert C, Double KL (2017) Evidence for reduced neurogenesis in the aging human hippocampus despite stable stem cell markers. Aging Cell 16(5):1195–1199. https://doi.org/10.1111/acel.12641
22. Harrison DE, Astle CM, Delaittre JA (1978) Loss of proliferative capacity in immunohemopoietic stem cells caused by serial transplantation rather than aging. J Exp Med 147(5):1526–1531
23. Roose H, Cox B, Boretto M, Gysemans C, Vennekens A, Vankelecom H (2017) Major depletion of SOX2+ stem cells in the adult pituitary is not restored which does not affect hormonal cell homeostasis and remodelling. Sci Rep 7(1):16940. https://doi.org/10.1038/s41598-017-16796-2
24. Metcalfe C, Kljavin NM, Ybarra R, de Sauvage FJ (2014) Lgr5+ Stem cells are indispensable for radiation-induced intestinal regeneration. Cell Stem Cell 14(2):149–159. https://doi.org/10.1016/j.stem.2013.11.008
25. Grégoire D, Kmita M (2014) Genetic cell ablation. In: Methods in molecular biology, Clifton, pp 421–436
26. Richardson WD, Young KM, Tripathi RB, McKenzie I (2011) NG2-glia as multipotent neural stem cells: fact or fantasy? Neuron 70(4):661–673. https://doi.org/10.1016/j.neuron.2011.05.013
27. Kuhn HG, Dickinson-Anson H, Gage FH (1996) Neurogenesis in the dentate gyrus of the adult rat: age-related decrease of neuronal progenitor proliferation. J Neurosci 16(6):2027–2033
28. Maslov AY, Barone TA, Plunkett RJ, Pruitt SC (2004) Neural stem cell detection, characterization, and age-related changes in the subventricular zone of mice. J Neurosci 24(7):1726–1733
29. Ahlenius H, Visan V, Kokaia M, Lindvall O, Kokaia Z (2009) Neural stem and progenitor cells retain their potential for proliferation and differentiation into functional neurons despite lower number in aged brain. J Neurosci 29(14):4408–4419. https://doi.org/10.1523/JNEUROSCI.6003-08.2009
30. Katsimpardi L, Litterman NK, Schein PA, Miller CM, Loffredo FS, Wojtkiewicz GR et al (2014) Vascular and neurogenic rejuvenation of the aging mouse brain by young systemic factors. Science 344(6184):630–634. https://doi.org/10.1126/science.1251141

31. Lichtenwalner RJ, Forbes ME, Bennett SA, Lynch CD, Sonntag WE, Riddle DR (2001) Intracerebroventricular infusion of insulin-like growth factor-I ameliorates the age-related decline in hippocampal neurogenesis. Neuroscience 107(4):603–613
32. Villeda SA, Plambeck KE, Middeldorp J, Castellano JM, Mosher KI, Luo J et al (2014) Young blood reverses age-related impairments in cognitive function and synaptic plasticity in mice. Nat Med 20(6):659–663. https://doi.org/10.1038/nm.3569
33. Brack AS, Conboy MJ, Roy S, Lee M, Kuo CJ, Keller C et al (2007) Increased Wnt signaling during aging alters muscle stem cell fate and increases fibrosis. Science 317(5839):807–810
34. Conboy IM, Conboy MJ, Wagers AJ, Girma ER, Weissman IL, Rando TA (2005) Rejuvenation of aged progenitor cells by exposure to a young systemic environment. Nature 433(7027):760–764
35. Carlson BM, Faulkner JA (1989) Muscle transplantation between young and old rats: age of host determines recovery. Am J Physiol 256(6 Pt 1):C1262–C1266
36. Harrison DE (1973) Normal production of erythrocytes by mouse marrow continuous for 73 months. Proc Natl Acad Sci U S A 70(11):3184–3188
37. Dykstra B, Olthof S, Schreuder J, Ritsema M, de Haan G (2011) Clonal analysis reveals multiple functional defects of aged murine hematopoietic stem cells. J Exp Med 208(13):2691–2703. https://doi.org/10.1084/jem.20111490
38. Liang Y, Van Zant G, Szilvassy SJ (2005) Effects of aging on the homing and engraftment of murine hematopoietic stem and progenitor cells. Blood 106(4):1479–1487
39. Morrison SJ, Wandycz AM, Akashi K, Globerson A, Weissman IL (1996) The aging of hematopoietic stem cells. Nat Med 2(9):1011–1016
40. Sudo K, Ema H, Morita Y, Nakauchi H (2000) Age-associated characteristics of murine hematopoietic stem cells. J Exp Med 192(9):1273–1280
41. Rossi DJ, Bryder D, Zahn JM, Ahlenius H, Sonu R, Wagers AJ et al (2005) Cell intrinsic alterations underlie hematopoietic stem cell aging. Proc Natl Acad Sci U S A 102(26):9194–9199
42. Chakkalakal JV, Jones KM, Basson MA, Brack AS (2012) The aged niche disrupts muscle stem cell quiescence. Nature 490(7420):355–360. https://doi.org/10.1038/nature11438
43. Sinha M, Jang YC, Oh J, Khong D, Wu EY, Manohar R et al (2014) Restoring systemic GDF11 levels reverses age-related dysfunction in mouse skeletal muscle. Science 344(6184):649–652. https://doi.org/10.1126/science.1251152
44. Ciccia A, Elledge SJ (2010) The DNA damage response: Making it safe to play with knives. Mol Cell 40(2):179–204. https://doi.org/10.1016/j.molcel.2010.09.019
45. Sahin E, Depinho RA (2010) Linking functional decline of telomeres, mitochondria and stem cells during ageing. Nature 464(7288):520–528. https://doi.org/10.1038/nature08982
46. Moskalev AA, Shaposhnikov MV, Plyusnina EN, Zhavoronkov A, Budovsky A, Yanai H et al (2013) The role of DNA damage and repair in aging through the prism of koch-like criteria. Ageing Res Rev 12(2):661–684. https://doi.org/10.1016/j.arr.2012.02.001
47. Olive PL, Banáth JP (2006) The comet assay: A method to measure DNA damage in individual cells. Nat Protoc 1(1):23–29
48. Jackson SP, Bartek J (2009) The DNA-damage response in human biology and disease. Nature 461(7267):1071–1078. https://doi.org/10.1038/nature08467
49. Rothkamm K, Barnard S, Moquet J, Ellender M, Rana Z, Burdak-Rothkamm S (2015) DNA damage foci: meaning and significance. Environ Mol Mutagen 56(6):491–504. https://doi.org/10.1002/em.21944
50. Panier S, Boulton SJ (2014) Double-strand break repair: 53BP1 comes into focus. Nat Rev Mol Cell Biol 15(1):7–18. https://doi.org/10.1038/nrm3719
51. Montpetit AJ, Alhareeri AA, Montpetit M, Starkweather AR, Elmore LW, Filler K et al (2014) Telomere length: a review of methods for measurement. Nurs Res 63(4):289–299. https://doi.org/10.1097/NNR.0000000000000037
52. Gordon LB, Rothman FG, López-Otín C, Misteli T (2014) Progeria: a paradigm for translational medicine. Cell 156(3):400–407. https://doi.org/10.1016/j.cell.2013.12.028

53. Gonzalo S, Kreienkamp R, Askjaer P (2017) Hutchinson-gilford progeria syndrome: a premature aging disease caused by LMNA gene mutations. Ageing Res Rev 33:18–29. https://doi.org/10.1016/j.arr.2016.06.007
54. Oshima J, Sidorova JM, Monnat RJ (2017) Werner syndrome: clinical features, pathogenesis and potential therapeutic interventions. Ageing Res Rev 33:105–114. https://doi.org/10.1016/j.arr.2016.03.002
55. Dokal I (2011) Dyskeratosis congenita. Hematol Am Soc Hematol Educ Progr 2011:480–486. https://doi.org/10.1182/asheducation-2011.1.480
56. Opresko PL, Shay JW (2017) Telomere-associated aging disorders. Ageing Res Rev 33:52–66. https://doi.org/10.1016/j.arr.2016.05.009
57. Karikkineth AC, Scheibye-Knudsen M, Fivenson E, Croteau DL, Bohr VA (2017) Cockayne syndrome: clinical features, model systems and pathways. Ageing Res Rev 33:3–17. https://doi.org/10.1016/j.arr.2016.08.002
58. Brosh RM, Bellani M, Liu Y, Seidman MM (2017) Fanconi anemia: a DNA repair disorder characterized by accelerated decline of the hematopoietic stem cell compartment and other features of aging. Ageing Res Rev 33:67–75. https://doi.org/10.1016/j.arr.2016.05.005
59. Blasco M, Lee HW, Hande MP, Samper E, Lansdorp PM, RA DP et al (1997) Telomere shortening and tumor formation by mouse cells lacking telomerase RNA. Cell 91(1):25–34
60. Rudolph KL, Chang S, Lee HW, Blasco M, Gottlieb GJ, Greider C et al (1999) Longevity, stress response, and cancer in aging telomerase-deficient mice. Cell 96(5):701–712
61. Baker DJ, Jeganathan KB, Cameron JD, Thompson M, Juneja S, Kopecka A et al (2004) BubR1 insufficiency causes early onset of aging-associated phenotypes and infertility in mice. Nat Genet 36(7):744–749
62. Chang S, Multani AS, Cabrera NG, Naylor ML, Laud P, Lombard D et al (2004) Essential role of limiting telomeres in the pathogenesis of Werner syndrome. Nat Genet 36(8):877–882
63. Yang SH, Bergo MO, Toth JI, Qiao X, Hu Y, Sandoval S et al (2005) Blocking protein farnesyltransferase improves nuclear blebbing in mouse fibroblasts with a targeted Hutchinson-Gilford progeria syndrome mutation. Proc Natl Acad Sci U S A 102(9):10291–10296
64. Yang SH, Meta M, Qiao X, Frost D, Bauch J, Coffinier C et al (2006) A farnesyltransferase inhibitor improves disease phenotypes in mice with a Hutchinson-Gilford progeria syndrome mutation. J Clin Invest 116(8):2115–2121
65. Osorio FG, Navarro CL, Cadiñanos J, López-Mejía IC, Quirós PM, Bartoli C et al (2011) Splicing-directed therapy in a new mouse model of human accelerated aging. Sci Transl Med 3(106):106ra107. https://doi.org/10.1126/scitranslmed.3002847
66. Villa-Bellosta R, Rivera-Torres J, Osorio FG, Acín-Pérez R, Enriquez JA, López-Otín C et al (2013) Defective extracellular pyrophosphate metabolism promotes vascular calcification in a mouse model of Hutchinson-Gilford progeria syndrome that is ameliorated on pyrophosphate treatment. Circulation 127(24):2442–2451. https://doi.org/10.1161/CIRCULATIONAHA.112.000571
67. Lebel M, Leder P (1998) A deletion within the murine werner syndrome helicase induces sensitivity to inhibitors of topoisomerase and loss of cellular proliferative capacity. Proc Natl Acad Sci U S A 95(22):13097–13102
68. Lebel M, Lavoie J, Gaudreault I, Bronsard M, Drouint R (2003) Genetic cooperation between the Werner syndrome protein and poly(ADP-ribose) polymerase-1 in preventing chromatid breaks, complex chromosomal rearrangements, cancer in mice. Am J Pathol 162(5):1559–1569
69. Labbé A, Garand C, Cogger VC, Paquet ER, Desbiens M, Le Couteur DG et al (2011) Resveratrol improves insulin resistance hyperglycemia and hepatosteatosis but not hypertriglyceridemia, inflammation, and life span in a mouse model for werner syndrome. J Gerontol A Biol Sci Med Sci 66(3):264–278. https://doi.org/10.1093/gerona/glq184
70. Massip L, Garand C, Turaga RVN, Deschênes F, Thorin E, Lebel M (2006) Increased insulin, triglycerides, reactive oxygen species, and cardiac fibrosis in mice with a mutation in the helicase domain of the Werner syndrome gene homologue. Exp Gerontol 41(2):157–168

71. Van der Horst GTJ, Van Steeg H, Berg RJW, Van Gool AJ, De Wit J, Weeda G et al (1997) Defective transcription-coupled repair in cockayne syndrome B mice is associated with skin cancer predisposition. Cell 89(3):425–435
72. Scheibye-Knudsen M, Ramamoorthy M, Sykora P, Maynard S, Lin P-C, Minor RK et al (2012) Cockayne syndrome group B protein prevents the accumulation of damaged mitochondria by promoting mitochondrial autophagy. J Exp Med 209(4):855–869. https://doi.org/10.1084/jem.20111721
73. Herrera E, Samper E, Martín-Caballero J, Flores JM, Lee HW, Blasco MA (1999) Disease states associated with telomerase deficiency appear earlier in mice with short telomeres. EMBO J 18(11):2950–2960
74. Lee HW, Gottlieb GJ, Horner JW, Greider CW, DePinho RA (1998) Essential role of mouse telomerase in highly proliferative organs. Nature 392(6676):569–574
75. Leri A, Franco S, Zacheo A, Barlucchi L, Chimenti S, Limana F et al (2003) Ablation of telomerase and telomere loss leads to cardiac dilatation and heart failure associated with p53 upregulation. EMBO J 22(1):131–139
76. Samper E, Fernández P, Eguía R, Martín-Rivera L, Bernad A, Blasco M et al (2002) Long-term repopulating ability of telomerase-deficient murine hematopoietic stem cells. Blood 99(8):2767–2775
77. Trifunovic A, Wredenberg A, Falkenberg M, Spelbrink JN, Rovio AT, Bruder CE et al (2004) Premature ageing in mice expressing defective mitochondrial DNA polymerase. Nature 429(6990):417–423
78. Sen P, Shah PP, Nativio R, Berger SL (2016) Epigenetic mechanisms of longevity and aging. Cell 166(4):822–839. https://doi.org/10.1016/j.cell.2016.07.050
79. Feser J, Truong D, Das C, Carson JJ, Kieft J, Harkness T et al (2010) Elevated histone expression promotes life span extension. Mol Cell 39(5):724–735. https://doi.org/10.1016/j.molcel.2010.08.015
80. O'Sullivan RJ, Kubicek S, Schreiber SL, Karlseder J (2010) Reduced histone biosynthesis and chromatin changes arising from a damage signal at telomeres. Nat Struct Mol Biol 17(10):1218–1225. https://doi.org/10.1038/nsmb.1897
81. Greer EL, Maures TJ, Hauswirth AG, Green EM, Leeman DS, Maro GS et al (2010) Members of the H3K4 trimethylation complex regulate lifespan in a germline-dependent manner in C. elegans. Nature 466(7304):383–387. https://doi.org/10.1038/nature09195
82. Jin C, Li J, Green CD, Yu X, Tang X, Han D et al (2011) Histone demethylase UTX-1 regulates C. elegans life span by targeting the insulin/IGF-1 signaling pathway. Cell Metab 14(2):161–172. https://doi.org/10.1016/j.cmet.2011.07.001
83. Eisenberg T, Knauer H, Schauer A, Büttner S, Ruckenstuhl C, Carmona-Gutierrez D et al (2009) Induction of autophagy by spermidine promotes longevity. Nat Cell Biol 11(11):1305–1314. https://doi.org/10.1038/ncb1975
84. Kaeberlein M, McVey M, Guarente L (1999) The SIR2/3/4 complex and SIR2 alone promote longevity in Saccharomyces cerevisiae by two different mechanisms. Genes Dev 13(19):2570–2580
85. Tissenbaum HA, Guarente L (2001) Increased dosage of a sir-2 gene extends lifespan in Caenorhabditis elegans. Nature 410(6825):227–230
86. Rogina B, Helfand SL (2004) Sir2 mediates longevity in the fly through a pathway related to calorie restriction. Proc Natl Acad Sci 101(45):15998–16003
87. Kanfi Y, Naiman S, Amir G, Peshti V, Zinman G, Nahum L et al (2012) The sirtuin SIRT6 regulates lifespan in male mice. Nature 483(7388):218–221. https://doi.org/10.1038/nature10815
88. Herranz D, Muñoz-Martin M, Cañamero M, Mulero F, Martinez-Pastor B, Fernandez-Capetillo O et al (2010) Sirt1 improves healthy ageing and protects from metabolic syndrome-associated cancer. Nat Commun 1:3. https://doi.org/10.1038/ncomms1001
89. Brown K, Xie S, Qiu X, Mohrin M, Shin J, Liu Y et al (2013) SIRT3 reverses aging-associated degeneration. Cell Rep 3(2):319–327. https://doi.org/10.1016/j.celrep.2013.01.005

90. Grewal SIS, Jia S (2007) Heterochromatin revisited. Nat Rev Genet 8(1):35–46
91. Pegoraro G, Kubben N, Wickert U, Göhler H, Hoffmann K, Misteli T (2009) Ageing-related chromatin defects through loss of the NURD complex. Nat Cell Biol 11(10):1261–1267. https://doi.org/10.1038/ncb1971
92. Tsurumi A, Li WX (2012) Global heterochromatin loss: a unifying theory of aging? Epigenetics 7(7):680–688. https://doi.org/10.4161/epi.20540
93. Larson K, Yan SJ, Tsurumi A, Liu J, Zhou J, Gaur K et al (2012) Heterochromatin formation promotes longevity and represses ribosomal RNA synthesis. PLoS Genet 8(1):e1002473. https://doi.org/10.1371/journal.pgen.100247
94. Narita MM, Nũnez S, Heard E, Narita MM, Lin AW, Hearn SA et al (2003) Rb-mediated heterochromatin formation and silencing of E2F target genes during cellular senescence. Cell 113(6):703–716
95. Laird PW (2010) Principles and challenges of genome-wide DNA methylation analysis. Nat Rev Genet 11(3):191–203. https://doi.org/10.1038/nrg2732
96. Cruickshanks HA, McBryan T, Nelson DM, Vanderkraats ND, Shah PP, van Tuyn J et al (2013) Senescent cells harbour features of the cancer epigenome. Nat Cell Biol 15(12):1495–1506. https://doi.org/10.1038/ncb2879
97. Maegawa S, Hinkal G, Kim HS, Shen L, Zhang L, Zhang J et al (2010) Widespread and tissue specific age-related DNA methylation changes in mice. Genome Res 20(3):332–340. https://doi.org/10.1101/gr.096826.109
98. Shumaker DK, Dechat T, Kohlmaier A, Adam SA, Bozovsky MR, Erdos MR et al (2006) Mutant nuclear lamin a leads to progressive alterations of epigenetic control in premature aging. Proc Natl Acad Sci 103(23):8703–8708
99. Osorio FG, Varela I, Lara E, Puente XS, Espada J, Santoro R et al (2010) Nuclear envelope alterations generate an aging-like epigenetic pattern in mice deficient in zmpste24 metalloprotease. Aging Cell 9(6):947–957. https://doi.org/10.1111/j.1474-9726.2010.00621.x
100. Horvath S (2013) DNA methylation age of human tissues and cell types. Genome Biol 14(10):R115
101. Chen BH, Marioni RE, Colicino E, Peters MJ, Ward-Caviness CK, Tsai PC, et al (2016) DNA methylation - based measures of biological age : meta - analysis predicting time to death. Aging (Albany) 8(9):1844–1865. doi: https://doi.org/10.18632/aging.101020
102. Ni Z, Ebata A, Alipanahiramandi E, Lee SS (2012) Two SET domain containing genes link epigenetic changes and aging in Caenorhabditis elegans. Aging Cell 11(2):315–325. https://doi.org/10.1111/j.1474-9726.2011.00785.x
103. Dang W, Steffen KK, Perry R, Dorsey JA, Johnson FB, Shilatifard A et al (2009) Histone H4 lysine 16 acetylation regulates cellular lifespan. Nature 459(7248):802–807. https://doi.org/10.1038/nature08085
104. Imai S, Armstrong CM, Kaeberlein M, Guarente L (2000) Transcriptional silencing and longevity protein Sir2 is an NAD- dependent histone deacetylase. Nature 403(6771):795–800
105. Mostoslavsky R, Chua KF, Lombard DB, Pang WW, Fischer MR, Gellon L et al (2006) Genomic instability and aging-like phenotype in the absence of mammalian SIRT6. Cell 124(2):315–329
106. McCormick MA, Mason AG, Guyenet SJ, Dang W, Garza RM, Ting MK et al (2014) The SAGA histone deubiquitinase module controls yeast replicative lifespan via sir2 interaction. Cell Rep 8(2):477–486. https://doi.org/10.1016/j.celrep.2014.06.037
107. Grillari J, Grillari-Voglauer R, Jansen-Dürr P (2010) Post-translational modification of cellular proteins by ubiquitin and ubiquitin-like molecules: role in cellular senescence and aging. Adv Exp Med Biol 694:172–196
108. Efeyan A, Comb WC, Sabatini DM (2015) Nutrient-sensing mechanisms and pathways. Nature 517(7534):302–310. https://doi.org/10.1038/nature14190
109. Chantranupong L, Wolfson RL, Sabatini DM (2015) Nutrient-sensing mechanisms across evolution. Cell 161(1):67–83. https://doi.org/10.1016/j.cell.2015.02.041

110. Fontana L, Partridge L, Longo VD (2010) Extending healthy life span—from yeast to humans. Science 328(5976):321–326. https://doi.org/10.1126/science.1172539
111. Choi IY, Piccio L, Childress P, Bollman B, Ghosh A, Brandhorst S et al (2016) A diet mimicking fasting promotes regeneration and reduces autoimmunity and multiple sclerosis symptoms. Cell Rep 15(10):2136–2146. https://doi.org/10.1016/j.celrep.2016.05.009
112. Di Biase S, Lee C, Brandhorst S, Manes B, Buono R, Cheng CW et al (2016) Fasting-mimicking diet reduces ho-1 to promote t cell-mediated tumor cytotoxicity. Cancer Cell 30(1):136–146. https://doi.org/10.1016/j.ccell.2016.06.005
113. Cheng CW, Villani V, Buono R, Wei M, Kumar S, Yilmaz OH, et al (2017) Fasting-mimicking diet promotes ngn3-driven β-cell regeneration to reverse diabetes. Cell 168(5):775–788.e12. doi: https://doi.org/10.1016/j.cell.2017.01.040
114. Wei M, Brandhorst S, Shelehchi M, Mirzaei H, Cheng CW, Budniak J et al (2017) Fasting-mimicking diet and markers/risk factors for aging, diabetes, cancer, and cardiovascular disease. Sci Transl Med 9(377):eaai8700. https://doi.org/10.1126/scitranslmed.aai8700
115. Harrison DE, Strong R, Sharp ZD, Nelson JF, Astle CM, Flurkey K et al (2009) Rapamycin fed late in life extends lifespan in genetically heterogeneous mice. Nature 460(7253):392–395. https://doi.org/10.1038/nature08221
116. Selman C, Tullet JM, Wieser D, Irvine EE, Lingard SJ, Choudhury AI et al (2009) Ribosomal protein S6 kinase 1 signaling regulates mammalian life span. Science 326(5949):140–144. https://doi.org/10.1126/science.1177221
117. Johnson SC, Rabinovitch PS, Kaeberlein M (2013) mTOR is a key modulator of ageing and age-related disease. Nature 493(7432):338–345. https://doi.org/10.1038/nature11861
118. Bitto A, Ito TK, Pineda VV, Letexier NJ, Huang HZ, Sutlief E et al (2016) Transient rapamycin treatment can increase lifespan and healthspan in middle-aged mice. Elife 5(pii):e16351. https://doi.org/10.7554/eLife.16351
119. Miller RA, Harrison DE, Astle CM, Fernandez E, Flurkey K, Han M et al (2014) Rapamycin-mediated lifespan increase in mice is dose and sex dependent and metabolically distinct from dietary restriction. Aging Cell 13(3):468–477. https://doi.org/10.1111/acel.12194
120. Jia K, Chen D, Riddle DL (2004) The TOR pathway interacts with the insulin signaling pathway to regulate C. elegans larval development, metabolism and life span. Development 131(16):3897–3906
121. Kapahi P, Zid BM, Harper T, Koslover D, Sapin V, Benzer S (2004) Regulation of lifespan in Drosophila by modulation of genes in the TOR signaling pathway. Curr Biol 14(10):885–890
122. Kaeberlein M, Powers RW, Steffen KK, Westman EA, Hu D, Dang N et al (2005) Regulation of yeast replicative life span by TOR and Sch9 response to nutrients. Science 310(5751):1193–1196
123. Wu JJ, Liu J, Chen E, Wang J, Cao L, Narayan N et al (2013) Increased mammalian lifespan and a segmental and tissue-specific slowing of aging after genetic reduction of mTOR expression. Cell Rep 4(5):913–920. https://doi.org/10.1016/j.celrep.2013.07.030
124. Franceschi C, Bonafè M, Valensin S, Olivieri F, De Luca M, Ottaviani E et al (2000) Inflamm-aging. An evolutionary perspective on immunosenescence. Ann N Y Acad Sci 908:244–254
125. Franceschi C, Capri M, Monti D, Giunta S, Olivieri F, Sevini F et al (2007) Inflammaging and anti-inflammaging: a systemic perspective on aging and longevity emerged from studies in humans. Mech Ageing Dev 128(1):92–105
126. Franceschi C, Campisi J (2014) Chronic inflammation (Inflammaging) and its potential contribution to age-associated diseases. J Gerontol A Biol Sci Med Sci 69:S4–S9. https://doi.org/10.1093/gerona/glu057
127. Ostan R, Bucci L, Capri M, Salvioli S, Scurti M, Pini E et al (2008) Immunosenescence and immunogenetics of human longevity. Neuroimmunomodulation 15(4-6):224–240. https://doi.org/10.1159/000156466
128. Youm Y-H, Kanneganti T-D, Vandanmagsar B, Zhu X, Ravussin A, Adijiang A et al (2012) The NLRP3 inflammasome promotes age-related thymic demise and immunosenescence. Cell Rep 1(1):56–68. https://doi.org/10.1016/j.celrep.2011.11.005

129. Youm Y-H, Grant RW, McCabe LR, Albarado DC, Nguyen KY, Ravussin A et al (2013) Canonical nlrp3 inflammasome links systemic low-grade inflammation to functional decline in aging. Cell Metab 18(4):519–532. https://doi.org/10.1016/j.cmet.2013.09.010
130. Loffredo FS, Steinhauser ML, Jay SM, Gannon J, Pancoast JR, Yalamanchi P et al (2013) Growth differentiation factor 11 is a circulating factor that reverses age-related cardiac hypertrophy. Cell 153(4):828–839. https://doi.org/10.1016/j.cell.2013.04.015
131. Villeda SA, Luo J, Mosher KI, Zou B, Britschgi M, Bieri G et al (2011) The ageing systemic milieu negatively regulates neurogenesis and cognitive function. Nature 477(7362):90–94. https://doi.org/10.1038/nature10357
132. Sinha M, Jang YC, Oh J, Khong D, Wu EY, Manohar R et al (2014) Restoring systemic gdf11 levels mouse skeletal muscle. Science 344(6184):649–652. https://doi.org/10.1126/science.1251152
133. Baker DJ, Childs BG, Durik M, Wijers ME, Sieben CJ, Zhong J et al (2016) Naturally occurring p16(Ink4a)-positive cells shorten healthy lifespan. Nature 530(7589):184–189. https://doi.org/10.1038/nature16932
134. Baar MP, Brandt RMC, Putavet DA, Klein JDD, Derks KWJ, Bourgeois BRM et al (2017) Targeted apoptosis of senescent cells restores tissue homeostasis in response to chemotoxicity and aging. Cell 169(1):132–147. https://doi.org/10.1016/j.cell.2017.02.031
135. Acosta JC, Banito A, Wuestefeld T, Georgilis A, Janich P, Morton JP et al (2013) A complex secretory program orchestrated by the inflammasome controls paracrine senescence. Nat Cell Biol 15(8):978–990. https://doi.org/10.1038/ncb2784
136. Wolf MJ, Adili A, Piotrowitz K, Abdullah Z, Boege Y, Stemmer K et al (2014) Metabolic activation of intrahepatic cd8 + t cells and nkt cells causes nonalcoholic steatohepatitis and liver cancer via cross-talk with hepatocytes. Cancer Cell 26(4):549–564. https://doi.org/10.1016/j.ccell.2014.09.003
137. Wang F, Flanagan J, Su N, Wang LC, Bui S, Nielson A et al (2012) RNAscope: a novel in situ RNA analysis platform for formalin-fixed, paraffin-embedded tissues. J Mol Diag 14(1):22–29. https://doi.org/10.1016/j.jmoldx.2011.08.002
138. Collado M, Gil J, Efeyan A, Guerra C, Schuhmacher AJ, Barradas M et al (2005) Tumour biology: senescence in premalignant tumours. Nature 436(7051):642. https://doi.org/10.1038/436642a
139. Pérez-Mancera PA, Young ARJ, Narita M (2014) Inside and out: the activities of senescence in cancer. Nat Rev Cancer 14(8):547–558. https://doi.org/10.1038/nrc3773
140. Fumagalli M, Rossiello F, Clerici M, Barozzi S, Cittaro D, Kaplunov JM et al (2012) Telomeric DNA damage is irreparable and causes persistent DNA-damage-response activation. Nat Cell Biol 14(4):355–365. https://doi.org/10.1038/ncb2466
141. Serrano M, Lin AW, McCurrach ME, Beach D, Lowe SW (1997) Oncogenic ras provokes premature cell senescence associated with accumulation of p53 and p16INK4a. Cell 88(5):593–602. https://doi.org/10.1016/S0092-8674(00)81902-9
142. Kuilman T, Peeper DS (2009) Senescence-messaging secretome: SMS-ing cellular stress. Nat Rev Cancer 9(2):81–94. https://doi.org/10.1038/nrc2560
143. Collado M, Blasco MA, Serrano M (2007) Cellular senescence in cancer and aging. Cell 130(2):223–233. https://doi.org/10.1016/j.cell.2007.07.003
144. Herbig U, Ferreira M, Condel L, Carey D, Sedivy JM (2006) Cellular senescence in aging primates. Science 311(5765):1257
145. Wang C, Jurk D, Maddick M, Nelson G, Martin-Ruiz C, Von Zglinicki T (2009) DNA damage response and cellular senescence in tissues of aging mice. Aging Cell 8(3):311–323. https://doi.org/10.1111/j.1474-9726.2009.00481.x
146. Dimri GP, Lee X, Basile G, Acosta M, Scott G, Roskelley C et al (1995) A biomarker that identifies senescent human cells in culture and in aging skin in vivo. Proc Natl Acad Sci 92(20):9363–9367

147. Baker DJ, Wijshake T, Tchkonia T, LeBrasseur NK, Childs BG, van de Sluis B et al (2011) Clearance of p16Ink4a-positive senescent cells delays ageing-associated disorders. Nature 479(7372):232–236. https://doi.org/10.1038/nature10600
148. Baker DJ, Childs BG, Durik M, Wijers ME, Sieben CJ, Zhong J et al (2016) Naturally occurring p16 Ink4a-positive cells shorten healthy lifespan. Nature 530(7589):184–189. https://doi.org/10.1038/nature16932
149. Chang J, Wang Y, Shao L, Laberge R-M, Demaria M, Campisi J et al (2016) Clearance of senescent cells by {ABT263} rejuvenates aged hematopoietic stem cells in mice. Nat Med 22(1):78–83. https://doi.org/10.1038/nm.4010
150. Roos CM, Zhang B, Palmer AK, Ogrodnik MB, Pirtskhalava T, Thalji NM et al (2016) Chronic senolytic treatment alleviates established vasomotor dysfunction in aged or atherosclerotic mice. Aging Cell 15(5):973–977. https://doi.org/10.1111/acel.12458
151. Farr JN, Xu M, Weivoda MM, Monroe DG, Fraser DG, Onken JL et al (2017) Targeting cellular senescence prevents age-related bone loss in mice. Nat Med 23(9):1072–1079. https://doi.org/10.1038/nm.4385
152. Hayflick L, Moorhead PS (1961) The serial cultivation of human diploid cell strains. Exp Cell Res 25:585–621
153. Eom YW, Kim MA, Park SS, Goo MJ, Kwon HJ, Sohn S et al (2005) Two distinct modes of cell death induced by doxorubicin: apoptosis and cell death through mitotic catastrophe accompanied by senescence-like phenotype. Oncogene 24(30):4765–4777
154. Karlseder J, Smogorzewska A, de Lange T (2002) Senescence induced by altered telomere state, not telomere loss. Science 295(5564):2446–2449
155. Gniadecki R, Hansen M, Wulf HC (2000) Resistance of senescent keratinocytes to UV-induced apoptosis. Cell Mol Biol (Noisy-le-grand) 46(1):121–127
156. Hampel B, Wagner M, Teis D, Zwerschke W, Huber LA, Jansen-Dürr P (2005) Apoptosis resistance of senescent human fibroblasts is correlated with the absence of nuclear IGFBP-3. Aging Cell 4(6):325–330
157. Ryu SJ, Oh YS, Park SC (2007) Failure of stress-induced downregulation of Bcl-2 contributes to apoptosis resistance in senescent human diploid fibroblasts. Cell Death Differ 14(5):1020–1028
158. Chen W, Kang J, Xia J, Li Y, Yang B, Chen B et al (2008) p53-related apoptosis resistance and tumor suppression activity in UVB-induced premature senescent human skin fibroblasts. Int J Mol Med 21(5):645–653
159. Pasillas MP, Shields S, Reilly R, Strnadel J, Behl C, Park R et al (2015) Proteomic analysis reveals a role for Bcl2-associated athanogene 3 and major vault protein in resistance to apoptosis in senescent cells by regulating ERK1/2 activation. Mol Cell Proteomics 14(1):1–14. https://doi.org/10.1074/mcp.M114.037697
160. Sharpless NE, Sherr CJ (2015) Forging a signature of in vivo senescence. Nat Publ Gr 15(7):397–408. https://doi.org/10.1038/nrc3960
161. Salminen A, Kauppinen A, Kaarniranta K (2012) Emerging role of NF-??B signaling in the induction of senescence-associated secretory phenotype (SASP). Cell Signal 24(4):835–845. https://doi.org/10.1016/j.cellsig.2011.12.006
162. Nelson G, Wordsworth J, Wang C, Jurk D, Lawless C, Martin-Ruiz C et al (2012) A senescent cell bystander effect: senescence-induced senescence. Aging Cell 11(2):345–349. https://doi.org/10.1111/j.1474-9726.2012.00795.x
163. Coppé J-PP, Patil CK, Rodier F, Krtolica A, Beauséjour CM, Parrinello S et al (2010) A human-like senescence-associated secretory phenotype is conserved in mouse cells dependent on physiological oxygen. PLoS ONE 5(2):e9188. https://doi.org/10.1371/journal.pone.0009188
164. Karaman MW, Herrgard S, Treiber DK, Gallant P, Atteridge CE, Campbell BT et al (2008) A quantitative analysis of kinase inhibitor selectivity. Nat Biotechnol 26(1):127–132. https://doi.org/10.1038/nbt1358

165. Burd CE, Sorrentino JA, Clark KS, Darr DB, Krishnamurthy J, Deal AM et al (2013) Monitoring tumorigenesis and senescence in vivo with a p16ink4a-luciferase model. Cell 152(1-2):340–351. https://doi.org/10.1016/j.cell.2012.12.010
166. Krishnamurthy J, Torrice C, Ramsey MR, Kovalev GI, Al-Regaiey K, Su L et al (2004) Ink4a/Arf expression is a biomarker of aging. J Clin Invest 114(9):1299–1307
167. Gruber HE, Ingram JA, Norton HJ, Hanley EN (2007) Senescence in cells of the aging and degenerating intervertebral disc: immunolocalization of senescence-associated beta-galactosidase in human and sand rat discs. Spine (Phila Pa 1976) 32(3):321–327
168. Geng Y-Q, Guan J-T, Xu X-H, Fu Y-C (2010) Senescence-associated beta-galactosidase activity expression in aging hippocampal neurons. Biochem Biophys Res Commun 396(4):866–869. https://doi.org/10.1016/j.bbrc.2010.05.011
169. Sousa-Victor P, Gutarra S, García-Prat L, Rodriguez-Ubreva J, Ortet L, Ruiz-Bonilla V et al (2014) Geriatric muscle stem cells switch reversible quiescence into senescence. Nature 506(7488):316–321. https://doi.org/10.1038/nature13013
170. Maskey RS, Kim MS, Baker DJ, Childs B, Malureanu LA, Jeganathan KB et al (2014) Spartan deficiency causes genomic instability and progeroid phenotypes. Nat Commun 5:5744. https://doi.org/10.1038/ncomms6744
171. Xue W, Zender L, Miething C, Dickins RA, Hernando E, Krizhanovsky V et al (2007) Senescence and tumour clearance is triggered by p53 restoration in murine liver carcinomas. Nature 445(7128):656–660. https://doi.org/10.1038/nature05529
172. Collado M, Serrano M (2006) The power and the promise of oncogene-induced senescence markers. Nat Rev Cancer 6(6):472–476. https://doi.org/10.1038/nrc1884
173. Demaria M, Ohtani N, Youssef SAA, Rodier F, Toussaint W, Mitchell JRR et al (2014) An essential role for senescent cells in optimal wound healing through secretion of PDGF-AA. Dev Cell 31(6):722–733. https://doi.org/10.1016/j.devcel.2014.11.012
174. Agostini A, Mondragõn L, Bernardos A, Martínez-Máñez R, Dolores Marcos M, Sancenõn F et al (2012) Targeted cargo delivery in senescent cells using capped mesoporous silica nanoparticles. Angew Chemie Int Ed Engl 51(42):10556–10560. https://doi.org/10.1002/anie.201204663
175. de Magalhães JP, Stevens M, Thornton D (2017) The business of anti-aging science. Trends Biotechnol 35(11):1062–1073. https://doi.org/10.1016/j.tibtech.2017.07.004

# Chapter 4
# Geroscience

**Isabel Arrieta-Cruz and Armando Luna-López**

**Abstract** Geroscience is an emerging discipline that examines the relationship between biological mechanisms of aging across different species with the goal of understanding the molecular and cellular pathways underlying age-related diseases. Geroscience is based upon finding connections between the so called "hallmarks of aging", a term that refers to stress adaptation, epigenetics, inflammation, macromolecular damage, metabolism, proteostasis, stem cells and regeneration as well as nutrient sensing to elucidate processes damaged in chronic diseases highly prevalent in older people. In this chapter, we tried to explain the origins of Geroscience, its relevance to the study of aging and its connection to disease, as well as to emphasize specific experimental findings that resulted from studies of animal models focused in replicating the physiopathological features of age-related diseases such as neurodegenerative and cardiovascular diseases, sarcopenia and osteoporosis, cancer, diabetes and frailty. Finally, we discussed some potential biomarkers suggested to improve the diagnosis or accelerate the identification of therapeutic targets in order to minimize the negative impact of chronic diseases during aging.

**Keywords** Geroscience · Biology of aging · Animal models · Age-related diseases

## 4.1 Introduction

The aging process has been studied from several perspectives including molecular, physiological, geriatric, epidemiological, economic, cultural, psychological and social approaches. Currently, the molecular mechanisms underlying aging have been delineated through the discovery of various molecular pathways such as inflammation, oxidative stress, protein misfolding, cancer progression, hormonal and endocrine disruptors, energy metabolism, cell degeneration and death, neuronal plasticity, neuronal transmission, gut microbiota, etc., in non-vertebrate models, e.g. yeasts, worms and flies as well as in vertebrate models, e.g. zebrafish and mice.

I. Arrieta-Cruz (✉) · A. Luna-López (✉)
Department of Basic Research, National Institute of Geriatrics, Mexico City, Mexico
e-mail: iarrieta@inger.gob.mx; aluna@inger.gob.mx



C. García-Peña et al. (eds.), *Aging Research - Methodological Issues*,
https://doi.org/10.1007/978-3-319-95387-8_4

These studies have contributed significantly to our understanding of the basis of age-related diseases [1].

Traditionally, the normal aging process or its pathophysiological conditions have been explored from a fragmentary and partial point of view. To mitigate this limited approach Dr. Gordon Lithgow and his scientific group from the Buck Institute for Research on Aging, have coined the term Geroscience and advocated the development of an emerging discipline also named Geroscience, as an approach for understanding the processes of normal aging and age-related diseases through an integrative and inclusive point of view encompassing complex interrelationships between the fields of basic and medical science [2]. In this chapter, we will discuss how the so-called hallmarks of aging are fundamentals to accelerate the diagnosis, prognosis and therapeutics of chronic diseases highly prevalent in the aging population.

## 4.2 Geroscience: The Beginning of a New Discipline

Dr. Felipe Sierra from the National Institute on Aging (NIA) of the United States of America (USA) has been a long-time advocate of the scientific discipline known as Geroscience, and together with other experts have consolidated a "Geroscience Interest Group" (GIG) that already lead to a first international summit meeting held in National Institutes of Health, Maryland on 2013. Among the main goals of this meeting were: (1) to propose a holistic view of aging to address age-related diseases as a group; (2) to share new findings in the field of the biology of aging and assess their impact on a single or several age-related diseases; (3) to find connections between basic areas of the research such as adaptation to stress, epigenetics, inflammation, macromolecular damage, metabolism, proteostasis, stem cells and regeneration, also known collectively as "hallmarks of aging"; (4) to recognize the most important dietary, genetic and pharmacological interventions in animals models and their contribution to lifespan and healthspan; (5) to develop new animals models of aging and, finally (6) promote comparative studies to detect new susceptibility factors for chronic diseases [3]. A consensus in the scientific community recognizes aging as an important risk factor for the development of chronic diseases. Following this premise, the GIG of the NIA organized a second Geroscience summit in order to discuss how three specific diseases -cancer, HIV/AIDS and diabetes mellitus (DM)- are driving aging, a new idea called "Reverse Geroscience". The main goal of this meeting was to explain whether or not chronic diseases are impacting the cellular mechanisms grouped as hallmarks of aging in a transitory or permanent manner [4].

## 4.3 Advances in Geroscience: Therapies and Interventions

The main contribution of Geroscience to the study of aging has been to take advantage of the knowledge generated by a multidisciplinary approach. Examples of subjects widely explored by this discipline are: (1) cancer progression.- the time-dependent accumulation of cellular damage as well as prolonged exposure to carcinogens or genotoxic agents increase the abnormal expression and activity of proteins involved in the control of the cell growth and proliferation, modifying cellular senescence and programmed cell death producing the progression of tumors [5]; (2) cardiovascular disease.- the surge of macromolecular damage and the increase in the generation of reactive oxygen species by mitochondria in cardiac cells release an large number of molecular factors that give rise to vascular inflammation, cell senescence, apoptosis, decrease of cell renewal, altered proteostasis, reduction of angiogenesis, adverse extracellular matrix remodeling, impaired nitric oxide metabolism with endothelial dysfunction, etc., affecting the correct function of the heart and vascular system [6]; (3) neurodegenerative diseases.- Alzheimer's disease (AD) and Parkinson's disease (PD) have been the most studied neurodegenerative diseases worldwide and highly associated with aging. Animals models have been relevant to understand some pathophysiological features observed in AD or PD [7, 8]; (4) sarcopenia and osteoporosis.- the older adults suffer skeletal and muscle disorders causing disability and loss of movement significantly affecting their daily activities. Aging alters bone formation, accelerates bone resorption, enhances sympathetic tone, changes the parathyroid/vitamin D axis, impairs renal function, increases the loss of muscle mass and reduces muscle strength, etc., all these changes could be explained by the presence of critical molecular determinants during aging: excess mitochondrial autophagy, metabolic dysfunction, accumulation of reactive oxygen species and cell senescence in the stem cell pool [9, 10]; (5) DM.- several epidemiological studies have suggested that DM has a role in the development of dementia. However, the factors linking DM with AD are largely unclear. Some of the risk factors proposed to play a role in the neurodegenerative process and the progression of the dementia of AD are: hyperglycemia, insulin resistance, oxidative stress, activation of inflammatory citokines and damage to the micro/macrovascular system; all these factors have also been observed in patients with DM [11, 12]; and (6) frailty.- this condition produces changes in the adaptive mechanisms that promote resilience and significantly reduces homeostatic capabilities while increasing sensitivity to stress; in consequence a critical number of signs and symptoms converge in the same people, these include: muscle weakness, slowness, low physical activity, weight loss and exhaustion –according the frailty phenotype. The biological systems most affected by this condition are: endocrine, inflammatory, muscle-bone, central and peripheral nervous and immune [13].

## 4.4 Biomarkers of Aging and Age-Related Diseases

Nine molecular biomarkers, known as "hallmarks of aging", have been identified, categorized and related to the aging process, and these are: genomic instability, telomere attrition, epigenetic alterations, loss of proteostasis, lack of nutrient sensing, mitochondrial dysfunction, cellular senescence, stem cells exhaustion and altered intercellular communication (please refer to chapter 3-Biomedical research in aging; for a complete description of the hallmarks of aging). These biomarkers could help to understand the variables that accelerate, delay or diminish the aging process [14]. In the following sections, we will briefly describe main biomarkers identified and proposed for discerning some chronic diseases associated to aging.

### *4.4.1 Biomarkers for Neurodegenerative Diseases*

Biomarkers suggested for AD are the loss of proteostasis and protein misfolding mainly observed in the beta-amyloid peptides and amyloid precursor protein producing amyloid plaques [15] or by an abnormal hyperphosphorylation of tau protein inducing the formation of neurofibrillar tangles in specific brain areas [16]. Double Strand *DNA* Breaks (DSB) is another interesting biomarker for AD, because in human brains of patients with AD a high content of DSB has been observed in comparison with normal patients [17]. On the other side, dysfunctional synaptic transmission is a way to evaluate cellular communication; consequently, the detection of neurogranin, a neuronal protein which participates in synaptic signaling through the regulation of calmodulin availability, could be one of the most interesting and newer promising biomarker for the diagnostic and prognosis of AD [18]. PD is characterized for the loss of dopaminergic neurons in brain areas called basal ganglia and substantia nigra, in addition to the presence of Lewy bodies, which are composed by abnormal deposits of a protein called alpha-synuclein. This protein has been proposed as a main biomarker in PD [19]. Other proteins proposed as biomarkers of PD are Parkin, DJ-1 and LRKK2, which are involved in the maintenance of membrane potential of the mitochondria. Any modification or changes in their molecular functions alters ATP synthesis and produces an aberrant assembly of complexes in the mitochondria thus generating a deficient cellular communication in neurons of the brain areas affected by PD [20, 21].

### *4.4.2 Biomarkers for Muscle Disorders*

Sarcopenia is characterized by a significant decrease in muscle mass and strength, as well as slow physical performance [22]. One of the biomarkers more studied in sarcopenia is myosin heavy chain protein, a protein involved in myogenesis [23].

The maintenance of muscle mass depends on the balance between apoptosis and the mechanisms of regeneration. Thus, some biomarkers proposed to assess apoptosis are caspase 3, apoptosis-inducing factor, Apaf1, Bax and DSB proteins [24]; while proteins for detecting muscular regeneration as Pax7 and Pax3 or for identifying early or late myogenesis as MyoD or Myf5 have been proposed to detect the loss of muscle mass [25]. On the other side, muscle satellite cells (SC) are stem cells specifically localized between basal lamina and sarcolemma of myofibrils, SC participate in the regeneration of muscle in response to injury and their number decreases with aging. For these reasons SC are being considered as a biomarker to detect sarcopenia [26]. Finally, mitochondrial dysfunction is a determinant factor in the physiopathological events of sarcopenia because the mitochondria contribute significantly to energy production for muscle function. The biomarkers identified in the mitochondrial dynamics include fusion (Mitofusins: Mfn1 and Mfn2 and OPA1), fission (Fis1, Mff y Drp1*) and* biogenesis (PGC-1α) proteins; all of which are affected in sarcopenia [27].

### 4.4.3 *Biomarkers for Bone Diseases*

Osteoporosis is a disease characterized by low bone mass and deterioration of bone tissue. This disease is very common in older adults and is the main cause of bone fracture. Recently, biomarkers for bone have been classified into two categories: (1) bone formation and (2) bone reabsorption. The first group includes bone-specific alkaline phosphatase, procollagen type I N-terminal propeptide and procollagen I carboxyterminal propeptide [28]. In the second group, we have: pyridinoline, deoxypyridinoline, C-telopeptide, N-terminal telopeptide and tartrate-resistant acid phosphatase [29]. A newer biomarker proposed to study bone formation is sclerostin, a protein produced by the osteocytes. High levels of sclerostin have been detected in patients with high bone mineral density suggesting that this protein could be an excellent biomarker for mature osteocytes [30].

## 4.5 Perspectives

For a decade, Geroscience has been a fashionable interdisciplinary field with a positive reception and growing acceptance within the international scientific community that concentrates all possible efforts from different disciplines to reduce the negative impact, disability and progression of the chronic diseases during all the stages of aging. Since the birth of Geroscience, biomedical specialists have engaged in the study of aging, focusing their attempts to advance knowledge in order to help people affected by chronic diseases. The popularity of Geroscience in the gerontological and geriatrics world has encouraged the establishment of international scientific networks such as the Geroscience Network, integrated by 18 institutions from

United States of America [31]. This network includes basic scientists, clinicians, and other health professionals with the goal of exchanging ideas in different workshops focused on the mechanisms underlying of the aging process. Another example is the Geroscience Center for Brain Health and Metabolism (GERO) recently created in Chile, where scientists from different disciplines are doing fundamental research in close association with clinical fields from neuroscience, molecular biology and genetics to develop interventions to slow aging and counteract age-related diseases. Lastly, since 2010, Mexico created a national network on aging, health and social development to connect scientists interested in the study of the aging from different disciplines such as basic science, clinical practice, epidemiological and social to integrate and consolidate their advances in favor of the aging people. In addition, an international scientific meeting focused on Geroscience was held in Mexico City on October 2016, with the goal of emphasizing for the value of studying aging from an integrative point of view.

**Acknowledgments** The authors thank Roger Gutiérrez-Juárez for his helpful comments and English proofreading.

## References

1. Tian X, Seluanov A, Gorbunova V (2017) Molecular mechanisms determining lifespan in short- and long-lived species. Trends Endocrinol Metab 28(10):722–734
2. Sierra F, Kohanski R (2017) Geroscience and the trans-NIH Geroscience Interest Group, GSIG. Geroscience 39(1):1–5
3. Kennedy BK, Berger SL, Brunet A, Campisi J, Cuervo AM, Epel ES, Franceschi C, Lithgow GJ, Morimoto RI, Pessin JE, Rando TA, Richardson A, Schadt EE, Wyss-Coray T, Sierra F (2014) Geroscience: linking aging to chronic disease. Cell 159(4):709–713
4. Kohanski RA, Deeks SG, Gravekamp C, Halter JB, High K, Hurria A, Fuldner R, Green P, Huebner R, Macchiarini F, Sierra F (2016) Reverse geroscience: how does exposure to early diseases accelerate the age-related decline in health? Ann N Y Acad Sci 1386(1):30–44
5. He S, Sharpless NE (2016) The impact of aging on cancer progression and treatment. In: Sierra F, Kohanski R (eds) Advances in geroscience. Springer International Publishing, Switzerland, pp 53–84
6. Chiao YA, Lakatta E, Ungvari Z, Dai DF, Rabinovitch P (2016) Cardiovascular disease and aging. In: Sierra F, Kohanski R (eds) Advances in geroscience. Springer International Publishing, Switzerland, pp 121–160
7. Kerchner GA, Wyss-Coray T (2016) The role of aging in Alzheimer's disease. In: Sierra F, Kohanski R (eds) Advances in geroscience. Springer International Publishing, Switzerland, pp 197–227
8. Andersen JK, Chinta S (2016) Parkinson's disease and aging. In: Sierra F, Kohanski R (eds) Advances in geroscience. Springer International Publishing, Switzerland, pp 229–255
9. Melov S, Rosen CJ (2016) Aging and the bone-muscle interface. In: Sierra F, Kohanski R (eds) Advances in geroscience. Springer International Publishing, Switzerland, pp 257–275
10. Glowacki J, Vokes T (2016) Osteoporosis and mechanisms of skeletal aging. In: Sierra F, Kohanski R (eds) Advances in geroscience. Springer International Publishing, Switzerland, pp 277–308

11. Arrieta-Cruz I, Gutiérrez-Juárez R (2016) The role of insulin resistance and glucose metabolism dysregulation in the development of Alzheimer's disease. Rev Investig Clin 68(2):53–58
12. Musi N, Bartke A (2016) Diabetes and aging. In: Sierra F, Kohanski R (eds) Advances in geroscience. Springer International Publishing, Switzerland, pp 355–376
13. Fried L, Ferrucci L (2016) Etiological role of aging in chronic diseases: from epidemiological evidence to the new geroscience. In: Sierra F, Kohanski R (eds) Advances in geroscience. Springer International Publishing, Switzerland, pp 37–51
14. López-Otín C, Blasco MA, Partridge L, Serrano M, Kroemer G (2013) The hallmarks of aging. Cell 153(6):1194–1217
15. Grimmer T, Riemenschneider M, Förstl H, Henriksen G, Klunk WE, Mathis CA, Shiga T, Wester HJ, Kurz A, Drzezga A (2009) Beta amyloid in Alzheimer's disease: increased deposition in brain is reflected in reduced concentration in cerebrospinal fluid. Biol Psychiatry 65(11):927–934
16. Blennow K, Dubois B, Fagan AM, Lewczuk P, de Leon MJ, Hampel H (2015) Clinical utility of cerebrospinal fluid biomarkers in the diagnosis of early Alzheimer's disease. Alzheimers Dement 11(1):58–69
17. Su JH, Deng G, Cotman CW (1997) Neuronal DNA damage precedes tangle formation and is associated with up-regulation of nitrotyrosine in Alzheimer's disease brain. Brain Res 774(1–2):193–199
18. Mattsson N, Insel PS, Palmqvist S, Portelius E, Zetterberg H, Weiner M, Blennow K, Hansson O, Alzheimer's Disease Neuroimaging Initiative (2016) Cerebrospinal fluid tau, neurogranin, and neurofilament light in Alzheimer's disease. EMBO Mol Med 8(10):1184–1196
19. Wang X, Yu S, Li F, Feng T (2015) Detection of α-synuclein oligomers in red blood cells as a potential biomarker of Parkinson's disease. Neurosci Lett 599:115–119
20. Truban D, Hou X, Caulfield TR, Fiesel FC, Springer W (2017) PINK1, Parkin, and mitochondrial quality control: what can we learn about Parkinson's disease pathobiology? J Parkinsons Dis 7(1):13–29
21. Thomas KJ, Mccoy MK, Blackinton J, Beilina A, van der Brug M, Sandebring A, Miller D, Maric D, Cedazo-Minguez A, Cookson MR (2011) DJ-1 acts in parallel to the PINK1/parkin pathway to control mitochondrial function and autophagy. Hum Mol Genet 20(1):40–50
22. Cruz-Jentoft AJ, Baeyens JP, Bauer JM, Boirie Y, Cederholm T, Landi F et al (2010) Sarcopenia: European consensus on definition and diagnosis: report of the European working group on sarcopenia in older people. Age Ageing 39(4):412–423
23. Balagopal P, Schimke JC, Ades P, Adey D, Nair KS (2001) Age effect on transcript levels and synthesis rate of muscle MHC and response to resistance exercise. Am J Physiol Endocrinol Metab 280(2):E203–E208
24. Cheema N, Herbst A, McKenzie D, Aiken JM (2015) Apoptosis and necrosis mediate skeletal muscle fiber loss in age-induced mitochondrial enzymatic abnormalities. Aging Cell 14(6):1085–1093
25. Brzeszczyńska J, Meyer A, Mcgregor R, Schilb A, Degen S, Tadini V, Johns N, Langen R, Schols A, Glass DJ, Roubenoff R, Ross JA, Fearon KCH, Greig CA, Jacobi C (2018) Alterations in the in vitro and in vivo regulation of muscle regeneration in healthy ageing and the influence of sarcopenia. J Cachexia Sarcopenia Muscle 9(1):93–105
26. Rotini A, Martínez-Sarrà E, Duelen R, Costamagna D, Di Filippo ES, Giacomazzi G, Grosemans H, Fulle S, Sampaolesi M (2018) Aging affects the in vivo regenerative potential of human mesoangioblasts. Aging Cell. https://doi.org/10.1111/acel.12714
27. Marzetti E, Calvani R, Cesari M, Buford TW, Lorenzi M, Behnke BJ, Leeuwenburgh C (2013) Mitochondrial dysfunction and sarcopenia of aging: from signaling pathways to clinical trials. Int J Biochem Cell Biol 45(10):2288–2301
28. Glover SJ, Eastell R, McCloskey EV, Rogers A, Garnero P, Lowery J, Belleli R, Wright TM, John MR (2009) Rapid and robust response of biochemical markers of bone formation to teriparatide therapy. Bone 45(6):1053–1058

29. Fardellone P, Séjourné A, Paccou J, Goëb V (2014) Bone remodelling markers in rheumatoid arthritis. Mediators Inflamm. https://doi.org/10.1155/2014/484280
30. Hay E, Bouaziz W, Funck-Brentano T, Cohen-Solal M (2016) Sclerostin and bone aging: a mini-review. Gerontology 62(6):618–623
31. Sierra F (2016) Moving geroscience into uncharted waters. J Gerontol A Biol Sci Med Sci 71(11):1385–1387

# Chapter 5
# Descriptive Studies in Clinical Gerontology and Geriatrics

**Mario Ulises Pérez-Zepeda, Lorena Jocabed Rocha Balcázar, and Miguel Germán Borda**

**Abstract** Descriptive studies can be the point of departure for more complex studies both in geriatrics and in gerontology. The ongoing need of information regarding new conditions highly prevalent in the older adults, begins with a thorough description of these phenomenon in addition to the magnitude of them in specific populations and its associations with other variables. This newly generated knowledge can aid the general public and in particular health professionals in identifying these emerging conditions and leave aside current stigma that attributes these problems only to aging. This chapter will describe the different types of descriptive studies, illustrating with current literature both in the field.

**Keywords** Descriptive studies · Epidemiology of aging · Case studies · Case series

## 5.1 Introduction

There is a group of research reports that derive from close observation of clinical issues (i.e., taken directly from daily practice), such as case studies and case series. The objective of this type of study is to observe and collect data about a given phenomenon in an individual or group of individuals [1]. Alternatively, and closer to epidemiology, are studies that look only to the frequency with which a phenomenon

M. U. Pérez-Zepeda (✉)
Department of Geriatric Epidemiology, National Institute of Geriatrics, Mexico City, Mexico
e-mail: mperez@inger.gob.mx

L. J. R. Balcázar
Internal Medicine Department, Local General Hospital Number 27,
Mexican Institute of Social Security, Mexico City, Mexico
e-mail: jocabed_3791@hotmail.com

M. G. Borda
Aging Institute, Faculty of Medicine, Pontificia Universidad Javeriana, Bogotá, Colombia

Centre for Age-Related Diseases, Stavanger University Hospital, Stavanger, Norway
e-mail: mmborda@gmail.com

 
C. García-Peña et al. (eds.), *Aging Research - Methodological Issues*,
https://doi.org/10.1007/978-3-319-95387-8_5

is present, most often illnesses in a fixed place and time, such as prevalence studies and ecological studies [2]. These studies may be derived either from new data (e.g., clinical observations or records) or from databases (e.g., national surveys or files from health systems). Depending on the number of subjects and the presence of a reference group (or all of the subjects exposed to risk of suffering from the condition), descriptive studies are divided into: case studies, case series, ecological, surveillance and prevalence studies. Only the latter contemplates taking an entire group of exposed subjects to learn the frequency with which the phenomenon of interest occurs in all of that population. They cannot yield data of causality, but they can be analytical and show associations; in addition to the description of new conditions (see Chap. 12 for Health Systems Research in Aging).

## 5.2 Case Study

This is the simplest design that can be made in clinical research where, from observation, an individual or clinical case is described, due to its novelty. Uncommon cases, not-well described conditions, strange outcomes, newly developed interventions; are examples of categories that are reported under the category of case study. In the case of studies of aging, an example is dementia, having a number of novel interventions described or different outcomes or features of this disease. Moreover, some illnesses are considered to be uncommon in older adults compared to other age groups (e.g., autoimmune disorders). For example, a recently diagnosed case of Conn syndrome in an adult 95 year of age is a rare case in its own right, although the illness itself is not rare. Additionally, some illnesses and syndromes are overlooked in the elderly though as obvious or undiagnosed. For example, in the case of frailty and oropharyngeal dysphagia that are widely known as important causes of several complications but set aside with lower relevance.

Due to the increasing number of subjects of extreme age, the report of cases of different illnesses is useful for increasing available knowledge about the different manifestations that an older adult could present with an illness that is rare for this age group and also how special and specific interventions for detecting and treat some geriatric syndromes that usually are overlooked could have an impact on better outcomes. Unlike prevalence studies, case studies may have a follow-up on the subject to learn the nature of the illness described; this aspect is shared in the case series design.

In the example chosen to illustrate a case study from the set of older adults, a case is described of systemic lupus erythematosus accompanied by primary biliary cirrhosis [3]. The illness in other age groups is not common, and much less so in older adults, so that in the description of the particular characteristics found, new knowledge from the case may be learned.

## 5.3 Case Series

This type of report groups similar cases that are presented during a fixed period of time. When similar cases of an illness are presented over a short time (days or weeks) they could raise an alert for an epidemic [2]. In the case of older adults, a case series could be a substitute for clinical cohorts that require a larger number of subjects and that are difficult to put together in these age groups. As with the case study, definitions are very important; if no standard and accepted definition is available, the cases should be described completely instead of using a newly developed tag [4].

Both in this type of reporting and in the case studies, it is usual to review the available literature on the topic that supports the conclusions arrived at through the case or cases presented. These narrative reviews; provide a general overview of the topic being studied and used terminology that eventually would lead to a standardized description of the problem.

The case series do not report only illnesses; they might also report on the development of a new intervention or therapy, or adverse effects of these, as in the example shown below. In that report, six cases of infections of the joints following injections in the knee joint are presented [5]. In addition to clinical cases, it is also possible that cases with a more social or gerontological focus could be reported.

## 5.4 Prevalence

This type of study gives valuable information about how representative a condition is within a given population. It provides the basis for formulating hypotheses and also for running strategies that could impact the health of the population, for example public health policies. To have external validity, prevalence studies must start with a representative sample of the population to which it refers. In the recent years, several countries have been interested in cross-sectional studies due to the necessity to know the current status of their older population, which have had a huge impact on their internal policies. In aging research, there are a number of studies that have representative samples of different countries; these studies will be further reviewed in the chapter about longitudinal studies and cohorts (Chap. 8).

However, representative samples are not always possible in the case of older adults, since there are subgroups within this age group that are clearly differentiated from the rest, having a higher frequency of problems or illnesses, and the higher the age the higher will be the proportion of a particular illness. Therefore, when doing prevalence studies in older adults, not only the time and place of taking data must always be specified, but also the age group being referred to and whether there is a sampling strategy to avoid a misleading increase or decrease in groups of very advanced age. Knowing the type of place where the sample of older adults was taken is also very relevant in this age group because it is not the same to determine the prevalence of geriatric syndromes in residents of a personal care home as in ambulatory adults, since a larger prevalence of geriatric syndromes would always be found in nursing homes than in the ambulatory population [6].

One of the most common problems the researcher finds when developing prevalence studies in the senior citizen population is the lack of widely accepted definitions of some of the conditions (see Chap. 2 for a discussion on taxonomy). Although this has been improving over time, as more consensuses are reached about the problems that concern the old, there is still a wide margin in the way in which geriatric processes are defined. One of the most common examples is frailty, where a multiplicity of definitions has been used, resulting in prevalence reports that differ enormously from one another, beyond the population variability itself [7].

Despite what these reports might contain arising from the records, many of the problems the researchers are trying to resolve might not be specified in the documents consulted, either because they are dealing with a recently defined condition or because the condition had a different name in earlier times. For example, in a 1990 article by Thorslund and collaborators, they were looking for the prevalence of protein energy malnutrition in older outpatient adults. However, the definitions used in this study make it almost impossible to compare the results with the current concepts. Moreover, some of the criteria used for this condition might today be considered as part of frailty (inflammatory markers and skin test anergy) or sarcopenia (calorie and protein malnutrition). Even though it is an excellent study, its usefulness is questionable when trying to translate this knowledge to the present [8], or when trying to have a clear picture of the dimension on malnutrition in that particular population. In addition to the previously mentioned term (frailty), there is no agreement on the definition of the term sarcopenia, and there are many definitions that might be found in the literature as well as with a number of other conditions [9].

However, as the world faces an increase in the older population accompanied by a core of knowledge generated by research in these issues, a more accurate picture is generated. Therefore, it is necessary to keep studying, measuring and tracking prevalence of the different geriatric conditions –and also of the well-known chronic diseases– due to its dynamism and importance. All with the aim of creating more and better interventions for prevention, diagnosis and treatment.

## 5.5 Other Descriptive Studies

Some texts include within this category what ecological and surveillance studies [2]. Ecological studies correlate phenomena in populations, not in individuals. In the example presented below, visits to an emergency room are correlated with environmental pollution; a positive correlation is found between the number of visits to an emergency ward and the level of pollution [10]. On the other hand, surveillance studies describe the appearance of a given phenomenon. In the example presented, a system for monitoring falls is implemented, which enables researchers to learn the frequency of falls within a specific community. While it might appear as a study of cohorts, in monitoring studies the goal is only to try to describe a phenomenon in terms of frequency without putting forward specific hypotheses of causal exposures [11, 12].

## 5.6 Summaries of Descriptive Studies in Aging Research

### 5.6.1 *Prevalence*

Sarcopenia is a problem that is increasingly identified among older adults. An algorithm was recently developed to identify this condition. The objective of a recent study was to determine the prevalence of sarcopenia in a group of senior citizens in Mexico City using the algorithm for sarcopenia of the European working group. A cross-sectional study was carried out on community-dwelling older adults, using a sample of 345 adults 70 years of age or older. With the goal of determining the presence of sarcopenia, muscular mass and strength were measured as well as physical performance. Muscular mass was measured by the circumference of the calf, muscular force by the strength of the grip, and physical performance by the speed of walking. The cutoff points used were as those suggested in the same European algorithm. A total of 116 (33.6%) of the subjects were found with sarcopenia, 75 (48.5%) of them women and 41 (27.4%) of them men, with higher prevalence in subjects 80 years of age or older (50.4%). Sarcopenic obesity was found in five subjects (1.4%), moderate sarcopenia in 21 patients (6%) and severe sarcopenia in 94 subjects (27.2%).

This study along many others on the topic of sarcopenia shows the actual difficulty that researchers and clinicians face when trying to define a recently described condition, even that a clear definition was used, this is only one of about one dozen of definitions of this condition. Therefore, and appropriate description of how this algorithm was implemented is expected to be in the manuscript [9, 13].

### 5.6.2 *Case Study*

#### 5.6.2.1 Case 1

A 69 years old woman without a family history of autoimmune illness, only a history of hypertension, presented with transient arthralgia and arthritis over the previous 4 years. Later, she showed sensitivity to light, malar rash, and diffuse discoid lesions on her trunk and face, for which she was seeking medical attention. In addition, she reported a loss of weight of approximately 3 kg over the previous 3 months. The physical examination showed only synovitis of the wrists. The laboratory results were as follows: hyperglobulinemia (20 g/L), lymphopenia (850 cells/mm3); with platelets, creatinine, and a general exam of urine normal. The immunology profile confirmed: positivity in the antinuclear antibodies with titer 1:400 and anti-dsDNA 115 IU /mL, with serum complement normal. The following antibodies were negative: anti-La, anti-cardiolipin, lupus anticoagulant, anti-SM, anti-RNP, anti-SCL-70, anticentromere, rheumatoid factor, and anti-citrullinated peptide. With the foregoing, it was concluded that systemic lupus erythematosus (presence of 6 from 11 criteria of the American College of Rheumatology for lupus) was present.

One year later she developed liver dysfunction. The abdominal examination revealed hepatosplenomegaly, liver function tests found double the normal values. The antimitocondrial antibodies were positive (1:164), with anti-E2 positive fraction. Serology tests for the hepatitis B and hepatitis C virus were negative. The findings were consistent with the diagnosis of primary biliary cirrhosis so that, for testing and staging, a liver biopsy was carried out, which corroborated the diagnosis and was classified as being in stage 1. Treatment was given with 600 mg per day of ursodeoxycholic acid, normalizing liver function tests in a month.

Late onset of systemic lupus erythematosus is relatively rare, with a frequency of between 12 and 18%. Although the autoimmune mechanisms behind the association of these two autoimmune disorders are not completely understood, there are few cases reported for either illness. Moreover, the new onset of these diseases in older adults is still very rare, so descriptions found in clinical settings could aid at identifying them in daily routine work [3].

#### 5.6.2.2 Case 2

A 77-year-old woman admitted to the emergency room for 8 months of solid dysphagia, which progressed to aphagia. These symptoms previously raised the need for several consultations with a primary care physician and several endoscopies without finding a cause. After that, she was also evaluated by a psychiatrist who discarded mental illness as an etiology for her symptoms. Her physical exam showed generalized hypotrophy and secondary functional limitation. The general appearance of the patient oriented the clinicians to a possible diagnosis of sarcopenia. Her cognitive status was normal according to the screening tests, but she had a positive screening for malnutrition and dysphagia. Blood tests showed severe hypoalbuminemia, and barium swallow test was abnormal. Consensus diagnoses among the Units of Geriatric, Gastroenterology and Psychiatry were, functional oropharyngeal dysphagia, severe protein energetic malnutrition and unstable functional decline. Thus, a multidisciplinary approach and treatment were started focused on the deglutition process, allowing the patient to achieve an improvement. She was discharged from the hospital with ambulatory physical, respiratory and speech therapy, nutritional supplements, fractioned semi-soft diet/five per day, outpatient control with geriatrics and nutrition and strong educational base for the patient and the family focused on good habits and safe ways of eating and feeding. In the following 2 months after discharge, the patient was eating pureed and low-consistency food by herself with an improvement in her functionality and nutritional status. This case with follow-up shows the importance of having in mind the existence of oropharyngeal dysphagia and the good impact that its multidisciplinary approach and treatment can have [14]. This whole case pointed to potential new condition: sarcopenic dysphagia.

### 5.6.3 Case Series

Intra-articular injections of corticosteroids and hyaluronic acid in the knee are widely practiced as a conservative treatment for osteoarthritis. However, there are related side effects, in particular with the corticosteroids, mainly from infections. The microorganism most commonly found is *Staphylococcus aureus*, with the occasional participation of other organisms, including other strands of staphylococci and anaerobic bacteria.

In this report, the median age of the group was 75 years (64–87 years). Most of the patients had significant comorbidity. Three of the patients were treated with corticosteroids, and the other three with hyaluronic acid. All of the patients arrived to the emergency room 1–5 days after the injection. The main manifestations were pain, swelling of the extremities involved and difficulty walking. None of the older adults had fever at first.

The physical examination of the six patients revealed inflammation of the affected knee and pain with its movement (active and passive). Analysis of the synovial fluid revealed a lightly elevated white cell count (mostly neutrophils) in four of the patients; only two had positive Gram stain. Most patients had elevated erythrocyte sedimentation rate and C-reactive protein.

The resultant bacteria in the synovial fluid culture were staphylococcus or streptococcus. One patient had a sterile culture, probably because of oral antibiotic treatment before being admitted. Antibiotic treatment with cefazolin was started. It was adjusted according to the results of the culture. All of the patients were submitted to surgical treatment. In four of the patients the surgical intervention was carried out within the first 24 h. One patient, 64 years of age, refused surgery during the first 2 weeks of hospitalization and was initially treated only with antibiotics. Given that the condition did not improve, the patient finally consented and was intervened.

Three of the six patients were submitted to more than one surgery. Four patients were treated with formal arthrotomy and two with arthroscopy. The intraoperative findings in all cases were synovial congestion and purulent material. One patient, man 86 years of age, was admitted to the intensive care unit following the operation after his second arthrotomy with progressive sepsis. Later he developed septic shock. He was again taken to the operating room for an urgent supracondylar amputation. He continued to deteriorate and finally died of septic shock.

Post-surgery, all the patients were treated in the same way: the antibiotics were administered for 4 to 6 weeks after surgery, the knees were immobilized for 3 days after surgery and were submitted to physiotherapy, drainage continued for several days. The median hospital stay was 22.5 days (9–40 days). The infection was finally resolved in five patients.

Intra-articular injection of the knee is not an inoffensive procedure and may be harmful and potentially fatal. In addition, its long-term benefit continues to be questionable. This report shows rich information on septic arthritis secondary to intra-articular injections. This could give valuable information to clinicians on what to expect or bring up ideas to make new studies about the topic in researchers [5].

### 5.6.4 Ecological Studies

The objective of this study was to investigate the effect of daily levels of air pollution (levels of carbon monoxide, nitrogen dioxide, ozone, sulfur dioxide and other particles with an aerodynamic profile ≤10 microns) in morbidity and correlation with the daily number of visits to the emergency rooms due to lower respiratory illness in persons older than 64 years of age in the city of Sao Paulo, Brazil between 1996 and 1998. Generalized Poisson additive regression models were used and adjusted for the long-term trend and the climate, the days of the week and the daily number of admissions. Ozone and sulfur dioxide were the pollutants associated with visits to emergency department in older adults. These results reinforce the idea that air contamination can promote adverse health outcomes in older adults.

In this example, the interaction between pollutants and health conditions seem to be related. Even though it provides with information, the descriptive nature of the study precludes from having any conclusion, and further studies with other designs could evidence if the association is true [10].

### 5.6.5 Surveillance Studies

Falls are the main cause of fatal and non-fatal injuries among older people in the United States. Despite the importance of injuries caused by falls, epidemiological studies of them among older persons have not identified either their causes or the means of preventing them. Therefore, a system of monitoring was established in the community in Miami Beach, Florida, as part of a study to evaluate falls among old people. A total of 1827 events of injury from falls were produced in this community between July 1985 and June 1986. More than 85% (1567) of the persons who fell received attention in an emergency room (the main source of information). The other cases were identified from one of the three sources utilized: reports from firemen, medical registers of hospitalized patients or from the medical report of the doctor who first provided primary care. Most of the falls (97%) were listed as accidental.

More than 100 persons sought medical assistance because of a fall every month. The moment of the injury was known by 68% (1244) of the persons who fell.

Seventy-four percent of these falls (921) occurred during daylight hours. Fifty-four percent of the falls (986) occurred in or near the house, and 38% of them recorded the specific place in the home where the fall took place: 42% occurred in the bedroom, 34% in the bathroom, 9% in the kitchen, 5% on the stairs, 4% in the living room and the other 6% in other areas. This monitoring system helped to clarify the causes of older adult's falls and identified the efforts of appropriate prevention. This study describes the different features of falls on a specific community of older adults. More than establishing inferences, the main goal of these studies is just to describe what to expect of a certain phenomenon [12].

## 5.7 Conclusions

While this type of studies does not allow for conclusions or inferences of causality, it is useful for describing a phenomenon in its early stages. As well, these studies are of low cost and with few ethical implications. As with other age groups, what ensures fidelity of information for older adults is planning the study properly, taking into account the peculiarities presented by a study on aging. As discussed previously a field still to be addressed is that of taxonomy, in order to have clear definitions as the initial step in the standardization of data in the aging research field.

## References


1. Gallin JI, Ognibene FP (2012) Principles and practice of clinical research, 3rd edn. Elsevier/Academic Press, Amsterdam\Boston
2. Grimes DA, Schulz KF (2002) Descriptive studies: What they can and cannot do. Lancet 359(9301):145–149. https://doi.org/10.1016/S0140-6736(02)07373-7
3. Hammami S, Chaabane N, Mahmoudi H, Bdioui F, Saffar H (2013) Late-onset systemic lupus erythematosus-associated primary biliary cirrhosis. Int Arch Med 6(1):3. https://doi.org/10.1186/1755-7682-6-3
4. Carey TS, Boden SD (2003) A critical guide to case series reports. Spine 28(15):1631–1634. https://doi.org/10.1097/01.BRS.0000083174.84050.E5
5. Shemesh S, Heller S, Salai M, Velkes S (2011) Septic arthritis of the knee following intraarticular injections in elderly patients: report of six patients. Isr Med Assoc J 13(12):757–760
6. Dartigues JF (2005) Methodological problems in clinical and epidemiological research on ageing. Rev Epidemiol Sante Publique 53(3):243–249
7. Collard RM, Boter H, Schoevers RA, Oude Voshaar RC (2012) Prevalence of frailty in community-dwelling older persons: a systematic review. J Am Geriatr Soc 60(8):1487–1492. https://doi.org/10.1111/j.1532-5415.2012.04054.x
8. Thorslund S, Toss G, Nilsson I, von Schenck H, Symreng T, Zetterqvist H (1990) Prevalence of protein-energy malnutrition in a large population of elderly people at home. Scand J Prim Health Care 8(4):243–248
9. Arango-Lopera VE, Arroyo P, Gutirrez-Robledo LM, Pérez-Zepeda MU (2012) Prevalence of sarcopenia in mexico city. Eur Geriatr Med 3(3):157–160. https://doi.org/10.1007/s00198-012-2091-x

10. Martins LC, Latorre MR, Cardoso MR, Goncalves FL, Saldiva PH, Braga AL (2002) Air pollution and emergency room visits due to pneumonia and influenza in São Paulo, Brazil. Revista de saude publica 36(1):88–94
11. Rothman KJ, Greenland S, Lash TL (2008) Modern epidemiology, 3rd edn. Wolters Kluwer Health/Lippincott Williams & Wilkins, Philadelphia
12. DeVito CA, Lambert DA, Sattin RW, Bacchelli S, Ros A, Rodriguez JG (1988) Fall injuries among the elderly. Community-based surveillance. J Am Geriatr Soc 36(11):1029–1035
13. Perez-Zepeda MU, Gutierrez-Robledo LM, Arango-Lopera VE (2013) Sarcopenia prevalence. Osteoporos Int 24(3):799. https://doi.org/10.1007/s00198-012-2091-x
14. Borda MG, Venegas-Sanabria LC, Puentes-Leal GA, García-Cifuentes E, Chavarro-Carvajal DA, Cano CA (2017) Oropharyngeal dysphagia in older adults: the well-known tale. Geriatr Gerontol Int 17(6):1031–1033. https://doi.org/10.1111/ggi.13012

# Chapter 6
# Qualitative Research in Gerontology and Geriatrics

**Fernando A. Wagner, Laurens G. Van Sluytman, Halaevalu F. Ofahengaue Vakalahi, and Chioma Nwakanma Wosu**

**Abstract** What makes qualitative research a unique approach is its ability to create in-depth knowledge about the topic under study. In this chapter, we first discuss qualitative research methods in general. Several approaches exist to conduct qualitative research and we examine how these methods can be used to study issues that affect older adults, highlighting the relative strengths and weaknesses of these methods. We present and discuss several data collection strategies: observation, in depth interviews, focal discussion groups, diaries and other registries. We also provide a few ideas on how to overcome typical challenges faced when using qualitative methods to study issues that affect older adults.

**Keywords** Qualitative research · In-depth interviews · Sampling strategies in qualitative research

## 6.1 Introduction

Qualitative research methods are approaches used to conduct research that are qualitative in nature. This statement begins to appear more interesting when we ask about the specific characteristics of qualitative research in comparison to quantitative research. Is it true that all research, whether quantitative or qualitative, holds similar characteristics and therefore it is not possible to distinguish their products…? We posit that all research shares similar goals of describing, explaining, and predicting phenomena, but the scope and methods of qualitative research are different from those used in quantitative research. Does that mean that these two forms of research are so different that they cannot be used together or cannot form any type of partnership? Here, we will disagree. While the goals and methods may be different, qualitative and

F. A. Wagner (✉) · C. N. Wosu
University of Maryland School of Social Work, Baltimore, MD, USA
e-mail: fernando.wagner@ssw.umaryland.edu; chiwosu@gmail.com

L. G. Van Sluytman · H. F. Ofahengaue Vakalahi
Morgan State University School of Social Work, Baltimore, MD, USA
e-mail: laurens.vansluytman@morgan.edu; halaevalu.vakalahi@morgan.edu



C. García-Peña et al. (eds.), *Aging Research - Methodological Issues*,
https://doi.org/10.1007/978-3-319-95387-8_6

quantitative research methods are often used together, sometimes simultaneously, and other times sequentially (see Chap. 10 on Mixed Methods). The first take-home message in this chapter is that quantitative and qualitative research may be used independently or interdependently to study issues relevant to older adults.

## 6.2 The Niche of Qualitative Research

What makes qualitative research a unique approach is its ability to create in-depth knowledge about the topic under study, partly based on its ability to conduct naturalistic inquiries, but also in its data analysis and reporting tradition of placing emphasis on narratives rather than merely numbers and figures. We can also cite the traditional research questions for which qualitative research is best suited —namely, when we are interested in describing or explaining the *how's* and *why's* of processes and complex behaviors. If a metaphor is adequate, we can say that qualitative researchers engage cultural humility [1] to study a problem through an iterative approach, increasingly gaining confidence in the nature of the phenomena they seek to understand. Moreover, a distinctive characteristic of qualitative research is that data collection is always conducted in consideration of the specific situation of the participant (e.g., meeting the participant where she or he is, adapting the duration and timing of interviews or other data collection strategies) and the participant as an expert in his/her lived experience.

Several approaches exist for conducting qualitative research and specifically with older adults. Creswell (2003) identifies the following methods: narrative, phenomenology, ethnography, case study, and grounded theory, and explains that both narrative and phenomenology are best suited for the study of individuals, while case studies and grounded theory can be used to explore processes, activities and events, while ethnography might be used to learn about culture and shared behaviors [2].

## 6.3 Sampling Alternatives

One of the most important ways to increase the usefulness of qualitative research is by having adequate sampling plans and procedures [3]. It has been established that qualitative research should not attempt to compete with quantitative studies in terms of statistical *representativity*. In fact, what makes qualitative research so attractive and useful is the selection of carefully chosen samples to examine the research question(s). While there are no definite or exclusive sampling procedures, Hertzog offers helpful information for pilot studies [4]. We discuss below a few ideas on how to attain a sample that is most useful. In addition, qualitative research has several choices when it comes to sampling, including: deviant case sampling, typical case sampling, maximum variation sampling, respondent-driven sampling (including snowball sampling), convenience sampling, negative-case sampling, and key-informant sampling.

The different targets and features of these sampling strategies are discussed by Namey and Trotter [5] and summarized in the table below Table 6.1.

**Table 6.1** Sample strategies commonly used in qualitative research

| Sampling strategy | Characteristics, Uses |
|---|---|
| Atypical cases | Best when the interest is in identifying specific characteristics that make some people special, or different (e.g., migrants, high-achievers, etc.) |
| Typical cases | The researcher attempts to find commonalities in the sample |
| Maximum variation | Cases are selected to represent opposite poles of a continuum (e.g., patients who are most and least willing to accept a given treatment). |
| Respondent-driven | Such as snow-ball samples, useful when it is difficult to identify and recruit participants (e.g., drug use cases), or to work with stigmatized populations (e.g., transgender, HIV patients, among others) |
| Convenience | Easiest approach, but researchers need to fight selection bias |
| Negative-cases | Useful to identify characteristics and factors that might decrease the usefulness of an intervention, for example. |
| Key-informants | Assumes certain people have access to valuable information, or impact groups and communities in particular ways. |

## 6.4 Data Collection Strategies

Data collection strategies must correspond to the goals for the study. All in all, the main task is to record as accurately as possible the physical setting, participant behaviors and interactions, including verbal and non-verbal exchanges [6, 7]. Qualitative researchers have several methods to choose when they are developing their data collection strategies, in this case, with older adults. These include but are not limited to observation (participant and non-participant), in-depth interviews, focus group discussions, as well as documents and artifacts analysis.

### *6.4.1 Observation*

Qualitative research can use a variety of observation strategies to collect data. The degree to which the researcher gets involved can vary, depending on the characteristics of what is being observed, and the relationship with those who are the protagonists. Participant observation takes into account the potential influence of the observer on what is being observed, as well as the potential influence of what is being observed on the observer. In any case, the quality of the documentation is of outmost importance. In prior times, ethnographers travelled with teams of artists that would draw images of objects, people, and context, wrote extensive diaries, and even collected specimens of vegetables and animals. There is no doubt that the technological revolution makes observation much easier, in that, today it is quite easy to record conversations and film entire sequences of events. However, we must note that regardless of the level of participation, researchers collecting data through observation must honor Human Subjects Protection principles. For example, no conversation is to be recorded for research purposes without explicit permission of those involved. This is critical in research with vulnerable populations including older adults (see Chap. 16).

We briefly mentioned earlier the importance of documentation. Field notes can help identify and reassess behaviors, situations, or elements that at first sight might have been overlooked, inadvertently. An additional consideration is that filed notes may help reduce researcher bias, or at least identify it.

### 6.4.2 In–Depth Interviews

It is no secret that in qualitative research, the skills and preparedness of the interviewer is paramount. Here, the researcher wants to be prepared to observe and record pieces of data that will enable her or him to develop a greater understanding of the interactions and communications that will occur. Interviews can be structured or semi-structured, depending on the degree to which the interviewer will follow a script with questions, the extent and nature of probes inviting for clarification or elaboration on a specific topic. Structured interviews may yield data that is easier to merge or compare to other interviews, but semi-structured interviews allow for a greater and deeper exploration of participant ideas, experiences, and emotions.

An interview guide must be prepared in advance, including complete information on Human Subjects Protection and a series of questions that can generate an informational conversation with the participant. These questions can be conceptualized as generating challenges that invite participants to become involved in a natural conversation. The prepared interviewer will be able to posit probes inviting the participant to expand on issues, clarify the meaning of statements, or even expand on a particular item. Hence, the role and contributions from the interviewer cannot be overstated.

### 6.4.3 Focal Discussion Groups

Traditionally referred to as 'focus groups' the term "focal discussion groups" has been suggested in community participatory studies, where participants reject being seen as "lab rats" or "research subjects" and more as people engaged in a discussion about a topic of interest [8, 9]. Thus, the main task of the facilitator of a focal discussion group concerns recruiting individuals who share an interest on a specific topic (the discussion cannot refer to a general topic, for the multiplicity of opinions would prevent arriving at consensus statements or at least identification of main themes). However, it should be mentioned that focal discussion groups are singularly useful in identifying how social interactions relate to people's ideas and decisions, as the researcher can observe, register, and then later analyze patterns of inter-relations associated with content, affect, and context during the discussions. No other data collection strategy offers such a unique opportunity to observe how interactions occur in relation to certain topics and the emergence of collective meanings. For example, a trained group facilitator will identify the changes in tone and interaction patterns when a controversial topic is introduced, including fear, excitement, or simply put, rejection.

### 6.4.4 Diaries, Documents and Other Registries

Another data collection strategy that may prove useful in qualitative research, especially with older adults, is the collection of diaries, documents and other systematic registries, such as correspondence logs, expense reports, and calendars, among others. These data collection methods can be used either on purpose or as a secondary source of information. If participants are asked to record information about daily activities, the researcher can specify particular targets of interest and even bring about preventive measures for potential biases by requesting, for example, that records are created at specific times and/or days of the week, or by indicating a specific time of the day when the record is to be created (e.g., when having lunch). Diaries, documents and other registry data can be obtained from written notes but are not limited to such notes because other forms of resources such as telephone messages, pictures, and drawings can be used. For example, a study can identify factors that impact food intake or diets at nursing homes.

## 6.5 Analysis of Qualitative Data

It has been established that qualitative data typically take the form of text, voice, and pictures, among others. Unlike quantitative research, often the data analysis process is carried out in qualitative research within the data collection process, in an iterative process that is reciprocally informational [10]. The data collection may be enhanced or adjusted based on preliminary findings, and the interpretation of the data may change as more data are collected. A key point is that data must be interpreted to create information and knowledge. Hence, the analysis of qualitative data requires a process that transforms the original observations into more systematically organized representations [11]. Here are some generic steps for the analysis of data discussed by Creswell [2]:(1) Data preparation (e.g., transcribe voice records), (2) Get a general sense of the data in a first pass, (3) Coding to create categories, (4) Generate a description (of the settings and the categories created in the prior step and identify themes), (5) Determine how the information will be represented, and, (6) Interpret the meaning of the data.

Creating categories is really a massive undertaking that merits further discussion here. Recall that the main goal of this process is to discover meaningful patterns in the data and/or develop a conceptual representation that is grounded in the data and speaks to the goals of the study. The challenge is to develop this process in such a way that builds credibility to the analysis, and hence coding is a process that must be documented. Several approaches to coding exist; however, the one discussed here is chosen given its relative simplicity and its ability to leave a "paper trail" (or computer files) for documentation [12].

- Once focus group sessions or interviews are transcribed, lines in the document are numbered to facilitate linking codes and other notes to the original text. Specialized qualitative analysis software is available and can be used to facilitate the process but cannot substitute the creative work of the research team.
- "A priori" codes can be created based on the research goals and expertise of the research team. "Emergent" codes are identified from the data as the analysis progresses. This differentiation underscores the iterative nature of the analysis.
- The Grounded Theory approach begins by creating "open or initial codes" that identify key elements of each transcribed line or phrase; then, creating "concepts" or more general codes; then, creating categories by integrating concepts and relating each of these categories to a theoretical model (axial coding), and then developing a narrative to explain the interconnection between these categories (selective coding).

## 6.6 Peculiarities of Qualitative Research in Gerontology and Geriatrics

Qualitative research uses an approach that is naturalistic and as close to participants and their context as possible. These characteristics allow studies to adapt to participants' needs. However, the specific characteristics of older adults, their immediate social network(s), and their caregivers, must be considered when attempting to develop a qualitative study with this population. Although healthcare access and other aspects including transportation and physical barriers are powerful determinants of service utilization, as well as for research participation, we will not discuss these issues here because they have been discussed in this book elsewhere (Chap. 12. Health Systems Research for Aging). Yet, other peculiarities need to be taken into consideration for qualitative research with older adults. As they may be specific to each approach and data collection method so we discuss them separately in the following paragraphs.

### 6.6.1 In–Depth Interviews

If properly invited in a culturally competent way, older adults tend to enjoy sharing memories and talking about their experiences. But they often do not like being rushed. This means that a qualitative researcher needs to be patient and respectful of an older participant, and able to skillfully ask probing questions to further explore specific areas in a way that will maximize the information collected without disrupting or disrespecting the participant. Ample time needs to be allotted for in-depth interviews. A problematic situation may arise when someone else transports an older participant to an interview, for the researcher may also need to take this person's time and availability into consideration. Other challenges may also include: hearing issues, recall problems, comprehension, fatigue, medications that may affect mood and alertness, language barriers and other problems.

### 6.6.2 Group Activities

Focal group discussions are efficient data collection methods for qualitative research focusing on generating collective meanings but the topic must be adequate for this method, as well as the characteristics of the participants. If interaction between participants is somewhat limited, the focal group discussion may not be the best approach. This can happen when language, hearing, and understanding issues are present. For this reason, it is even more important than ever to ensure an adequate physical space for any qualitative research initiative involving groups of older adults.

Another important consideration is that the facilitator must be able to keep the discussion following a natural flow, such that conversations consistently return to the home theme, even though natural digressions may occur. The facilitator must also actively engage members who may experience reluctance in discussing potentially sensitive issues in a group. In sum, the group facilitator must keep in mind participants' needs for communication, along with the goal of making sure that the research questions set for the group are discussed and that collective meanings are generated.

### 6.6.3 Observation

Several inputs about everyday challenges and opportunities for older adults can be gained through observational studies of the type of ethnological research. Many questions about family relations, physician-patient interactions, and challenges to older adults can only be identified and explained through rigorous, systematic observational studies where researchers 'shadow' participants through their every-day activities. The researcher here will be continuously presented with the challenge of realizing how much closeness or distance he should keep. The issue is that it is almost impossible to perform the role of a completely unengaged observer, but even more challenging is the fact that the active-participant researcher will necessarily transform what would have naturally occurred had she/he not been involved.

## 6.7 Limitations of Qualitative Research

Qualitative research methods have important limitations that can serve as future directions. These include (1) the scope of qualitative research as an exercise that seeks to produce in-depth knowledge rather than representativeness; (2) in addition, qualitative studies' data processing and analyses require extensive work, especially in the preferred scenario that more than one researcher is involved in the analysis of data to increase the credibility and trustworthiness of the study; and, (3) the fact that qualitative methodologies usually do not seek to collect quantitative data to assess the number or proportion of people in a population who have a particular characteristic or who benefit from a specific program, which in many instances is the main concern of program funders and policy makers.

Perhaps the most important limitation of qualitative research is the need to trade-off depth over breadth [2, 5]. It is true that qualitative methods allow for much more flexibility than quantitative methods but collecting and analyzing qualitative data can take an enormous amount of effort, resources, and time, and therefore samples in qualitative research tend to be small, and not necessarily representative of a population. However, representativeness is also a challenging topic for quantitative research, with its significant issues in areas such as participation, accuracy, and ability to retain and follow subjects over time. Indeed, qualitative research commits to working with individuals and organizations and invests considerable time and energy in establishing and maintaining a strong relationship with participants. Yet, it can be argued that this is one of the main threats to the validity of qualitative research, the fact that it is too close to the participants to keep the necessary 'distance' to avoid compromising objectivity.

## 6.8 Enhancing Rigor in Qualitative Research

Because results of qualitative research are so intrinsically dependent on the researcher's ability to record unbiased data, several suggestions are offered here to increase the likelihood of adequate data and data analysis procedures.

One of the critical aspects of qualitative research is checking the accuracy of the findings. There are many ways in which researchers can use, including triangulation, request confirmation from participants (when possible), self-disclose potential biases of the research team, the meaning making process through team discussion of findings, discussion of discrepant information, submit data and findings for evaluation by independent referees [13].

## References


1. Ortega R, Coulborn K (2011) Training child welfare workers from an intersectional cultural humility perspective: a paradigm shift. Child Welfare 90(5):27–49
2. Creswell JC (2003) Research design. In: Qualitative, quantitative, and mixed methods approaches, 2nd edn. Sage Publications, Inc., Thousand Oaks
3. Green J, Thorogood N (2005) Qualitative methods for health research. Sage Publications, Inc., Thousand Oaks
4. Hertzog MA (2008) Considerations in determining sample size for pilot studies. Res Nurs Health 31(2):180–191. https://doi.org/10.1002/nur.20247
5. Namey EE, Trotter RT (2015) Qualitative research methods. In: Guest G, Namey EE (eds) Public health research methods. SAGE Publications, Inc., Thousand Oaks, pp 443–482
6. Agar M (1980) The professional stranger. Academic Press, San Diego
7. DiCicco-Bloom B, Crabtree BF (2006) The qualitative research interview. Med Educ 40(4):314–321
8. Wagner FA, Sheikhattari P, Buccheri J, Gunning M, Bleich L, Schutzman C (2016) A community-based participatory research on smoking cessation intervention for urban communities. J Health Care Poor Underserved 27(1):35–50
9. Sheikhattari P, Apata J, Kamangar F, Schutzman C, O'Keefe A, Buccheri J, Wagner FA (2016) Examining smoking cessation in a community-based versus clinic-based intervention using community-based participatory research. J Community Health 41(6):1146–1152
10. Shepherd SK & Acterberg CL (1992). Qualitative research methodology: data collection, analysis, interpretation, and verification. In E. Monsen (D.), Research. Successful approaches (pp. 82–99). Chicago: American Dietetic Association
11. Bernard HR, Ryan GW (2010) Analyzing qualitative data: systematic approaches. Sage Publications, Inc., California
12. Ulin PR, Robinson ET, Tolley EE (2005) Qualitative methods in public health. Jossey-Bass, San Francisco
13. Royse D, Thyer BA, Padgett DK (2016) Program evaluation: an introduction to an evidence-based approach, 6th edn. Wadsworth Cengage Learning, Belmont

# Chapter 7
# Case-Control Studies in Aging Research

**Sergio Sánchez-García, Erika Heredia-Ponce, Luis Pablo Cruz-Hervert, Ángel Cárdenas-Bahena, Laura Bárbara Velázquez-Olmedo, and Carmen García-Peña**

**Abstract** In the planning phase of research related to age and aging, the quality of knowledge derived from epidemiological studies, depends heavily on the solidity of the methodological design and the strategies for collecting data designed to answer the research question. The classic design of case-control studies distinguishes between older adults who have a specific outcome or disease (cases) and those who do not (controls) and are determines whether the subjects were exposed or not to one or several factors to try to establish, retrospectively (that is, from the effect to the cause), the relationship of these factors with the disease. Case-control studies are a cost-effective alternative for providing a valid and reasonably precise estimate for identifying an association force of a hypothetical relationship cause-effect in studies related to older adults. Recently case-control studies have been related directly to cohort studies, which enabled researchers to design new patterns for their development while obtaining major benefits. The case-control study is the appropriate choice and at times the only alternative for studying diseases of very low incidence in older adults. With this type of study, it is possible to explore a broad range

S. Sánchez-García (✉) · Á. Cárdenas-Bahena
Research Unit in Epidemiology and Health Services, Aging Area, National Medical Center Century XXI, Mexican Institute of Social Security, Mexico City, Mexico
e-mail: sergio.sanchezga@imss.gob.mx

E. Heredia-Ponce
Department of Public Health and Oral Epidemiology, Faculty of Dentistry, National Autonomous University of Mexico, Mexico City, Mexico
e-mail: ehp@unam.mx

L. P. Cruz-Hervert
Center for Research on Infectious Diseases, National Institute of Public Health, Mexico City, Mexico
e-mail: lhervert@insp.mx

L. B. Velázquez-Olmedo
Faculty of Dentistry, National Autonomous University of Mexico, Mexico City, Mexico

C. García-Peña
Research Division, National Institute of Geriatrics, Mexico City, Mexico
e-mail: mcgarcia@inger.gob.mx



C. García-Peña et al. (eds.), *Aging Research - Methodological Issues*,
https://doi.org/10.1007/978-3-319-95387-8_7

of related exposures to illness. Another important advantage is that they require smaller samples and are less costly than experimental designs or cohort studies. Among the disadvantages or drawbacks of this type of study is that it can only provide information about the event or target disease in the population that has been selected for carrying out the study.

**Keywords** Case-control studies · Cause-effect · Low-incidence studies

## 7.1 Introduction

Case-control studies are considered an achievement of modern epidemiology; they are defined as a non-experimental epidemiological analytical method for identifying potential causal relationships between the presence of a specific event or disease and previous exposures [1, 2]. A case-control study is called for when the research question addresses low-frequency events or diseases [3] such as Creutzfeldt-Jakob disease in older adults [4], or diseases that have long latency periods such as prostate cancer [5], as well as diseases with multiple potential etiological factors such as fall-related fractures [6]. A case-control study is also useful when a prospective cohort study is not feasible due to the required follow-up time (Chap. 8), or when research subjects are lost, which in the case of older adults is usually due to death the vast majority of times.

These study designs can also be used in health problems where a relatively fast approach is required and when the exposure or prospective procedure may be unethical, expensive or due to greater loss of research subjects. For example, in the study of the therapeutic use of bisphosphonates for the treatment of osteoporosis and its association with an increased risk of ischemic stroke [7].

Therefore, case-control studies are a cost-effective alternative to provide a valid and reasonably accurate estimate to identify a strength of association of a hypothetical cause-effect relationship [8].

## 7.2 Case-Control Studies and Design Variants

The classic design of case-control studies distinguishes between older adults who have a specific outcome or disease (cases) and those who do not (controls); the latter are, ideally, a random sample of the population from which the cases are drawn. It then determines whether the subjects were exposed or not to one or several factors to try to establish, retrospectively (that is, from the effect to the cause), the relationship of these factors with the disease [1, 9]. Figure 7.1 shows the design of a case-control study.

Recently, the possibility of combining the strategies of case-control studies and cohort studies has been considered (see Chap. 8 for a detailed description of cohort studies). This has led to new study designs that have shown a number of advantages, mainly regarding the cost of performing the study [8, 10].

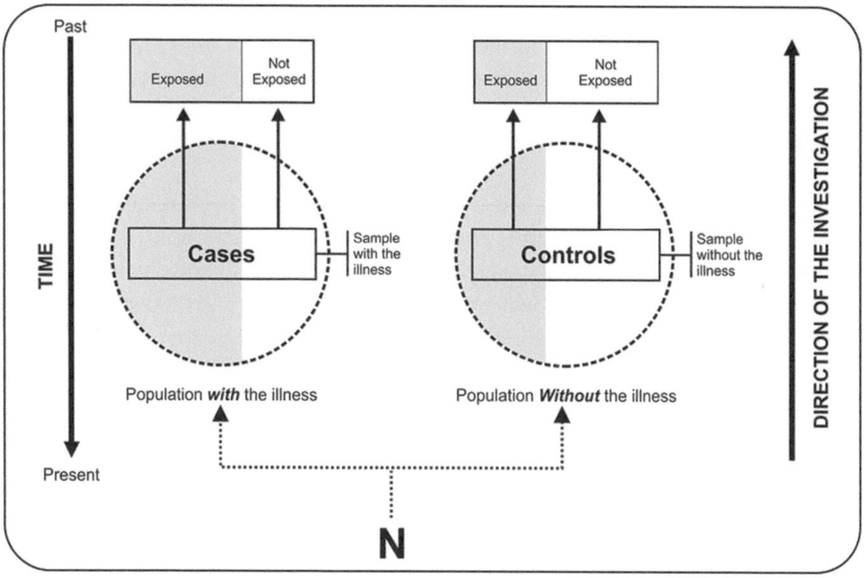


**Fig. 7.1** Classical design of a case-control study
*Left arrow*: TIME; past, present. *Right arrow*: DIRECTION OF THE INVESTIGATION
*Cases*: Exposed; Non-exposed; Sample with the illness, Population with the illness
*Controls*: Exposed; Non-exposed; Sample without the illness; Population without the illness

A number of variants or hybrid studies have been proposed, such as case-cohort studies, nested case-control studies, case-case studies, and proportional mortality studies.

In case-cohort studies, cases and controls belong to a specific cohort defined in time and space. This allows to estimate the cumulative incidence ratio, since all members of the cohort have the same follow-up time and the controls are randomly selected from the cohort with which the study was started [10, 11].

An example of this type of design is a study performed on a series of 100 consecutive patients subjected to colon (n = 44) or rectal (n = 56) resection in order to evaluate the postoperative results of the intravenous administration of fluids and of the administration of sodium after colon or rectal resection [11]. This type of study is also discussed in Chap. 8.

Nested case-control studies use a sampling scheme known as risk group, in which the case is selected from the exposed group. This design is recommended for the study of rare diseases in dynamic cohort studies in which determining the exposure and changes over time for all members of the cohort would be very expensive. An example of this type of study was conducted in Denmark with medical records from 1989 to 2008 to evaluate the effect of treatment with metformin on the risk of breast cancer among peri and postmenopausal women with diabetes mellitus type II [12].

Case-case studies are useful to determine what the subject was exposed to, in an unusual way, before the onset of the disease or event under study. In this type of design, individuals serve as their own control. An example of this design is a study of the relationship between an acute myocardial infarction and the occurrence of an episode of intense anger 2 h before the occurrence of the infarction event, compared with two paired self-controls [8, 13].

In proportional mortality studies, the cases are defined by deaths that occurred in a population source, while controls consist in deaths that occur in the base population but that are not related to exposure. A study conducted by Freedman et al. used death certificates to study whether mortality from multiple sclerosis is negatively associated with exposure to sunlight [14].

## 7.3 Sample Size

Since we start by knowing the subject who had the specific event or disease, the important thing is to establish, from the literature, the percentage of cases in which there was previous exposure, as well as the percentage with which exposure is present in the controls. Both percentages can then be compared and analyzed to determine the odds ratio or cross products ratio [15–18].

If this information is not available, it can be assumed that the percentage of cases with exposure is 50% and a plausible criterion or a clinically important difference can be used to estimate whether the proportion of exposure is lower (if it prevents) or higher (if it promotes the disease) with respect to the disease or event (cases). Likewise, it is necessary to establish a statistical significance value (alpha) of 0.05 and a statistical power (1-beta) of 0.80 in order to attain statistical validity [16].

In relatively infrequent diseases for which the number of available cases is limited, the statistical power of the study can be increased by using multiple controls (two or three for each case); however, when the ratio between controls and cases exceeds a value of 4:1, the gain in terms of statistical power becomes very small [10].

## 7.4 Selection of Cases

To select the cases, it is first necessary to have an operational definition of the event or disease that will be studied in order to properly identify the cases of interest [17]. An example of this type of operational definition can be found in the study of sarcopenia in older adults by Baumgartner et al., who defined it as a value equal or lower to two standard deviations from the mean of appendicular muscle mass in a representative population of young adults (18–40 years of age) [18].

An operational definition of a disease or event can also be based on a consensual definition. For example, it is possible to use the diagnostic criteria of the Diagnostic and Statistical Manual of Mental Disorders of the American Psychiatric Association, Fourth Edition to define cases of major depressive disorder in older adults [19].

Once the operational definition of the disease or event has been decided, there are different alternatives for selecting cases. One of the best methods is to include cases only when the event occurs or when the disease of interest is diagnosed, a strategy called "incident cases". This selection method is similar to a cohort study, since the cases are representative of all events or diseases of interest in the population from which they were obtained [10].

Another strategy is to select older adults already suffering from the disease or event of interest; this method is called "prevalent cases." A disadvantage of this method is that the selected episodes will be those with higher survival odds and the risk factors of these older adults may not represent those of the total population of older adults with the disease or event of interest. Another drawback is the risk of memory biases and changes in the history of exposure, including those that arise from the clinical course of the disease itself. Furthermore, this selection strategy makes it difficult to identify the population from which the cases originate and from which the controls will subsequently be obtained [8].

If one has access to the notification records for a certain event or disease in a particular population, it is possible to obtain cases from that database, provided that the selection criterion is specified. The common procedure is to take a sample and not all of the reported cases, except in the case of rare events or diseases. In order to gather enough cases of a rare disease or event, it is necessary to use both incident and prevalent cases, but they must be analyzed beforehand to ensure that both types of cases come from the same population; otherwise, they could lead to wrong conclusions [9].

It is possible to implement a strategy for selecting cases in a hospital, as long as it can be guaranteed that all or most of the older adults with the event or disease under study are treated and diagnosed in a particular hospital and the population to which they belong can be identified [20].

An example of a selected sample of "hospital cases" was reported by Suzuki et al., who studied the risk factors for hip fractures in older adults in Japan. In that study, most cases of hip fracture were diagnosed and treated in 21 hospitals that were included in the study because they covered 80% of the Japanese population in the year of 1989 [21].

The validity of case-control studies depends on whether the cases included adequately represent the history of exposure of all existing cases, thus avoiding selection biases, and on the proper registration of exposure antecedents, which allows to avoid information biases.

## 7.5 Selection of Controls

Before selecting the controls, it is necessary to identify the population from which the cases came, which is not always an easy task. The population can be defined by all those older adults who would have been considered as cases if they had developed the event or disease of interest [9].

The selection of controls must be done independently of their condition as exposed or unexposed in order to ensure that they adequately represent the population of origin. The latter can be achieved as long as the exposure condition does not determine the possibility that an older adult is included in the study as a control. This implies that the sample fractions should be the same for exposed and unexposed controls; however, in most of these studies, the sample fractions are unknown [8, 9].

The probability of a subject to be selected as control must be proportional to the time he or she remained eligible to develop the event or disease under study. That is, if an older adult migrated or died during the study, it should no longer be eligible to be selected as control. One way to account for this is by selecting a control from the eligible group each time a case is detected or selected; this is known as risk-group selection. This selection scheme ensures that the controls are at risk of developing the event at the time they are selected.

This scheme also implies that an older adult selected as a control at an early stage of the study could also be selected as a case in later stages of the study [8].

## 7.6 Biases

Case-control studies can be subject to various systematic errors, also known as biases. Such biases may be present in the procedures used to select and classify the cases and controls. Although researchers have a great interest in detecting biases, they often remain undetected.

One the most common types of bias in case-control studies is selection bias. As is known, the controls must be selected from the same population as the cases so that they can be said to represent a certain population. But this is not always true, as sometimes, even though the cases and controls belong to the same population, the characteristics of the cases differ from those of the general population (controls) [22, 23].

Another type of bias occurs when one of the two groups is more strictly observed at the time of capturing the results, giving rise to an observation bias. Information bias can occur in two ways: as memory limitations and as memory bias. The first is due to the variable limits of the ability to remember information, while memory bias has its cause in the non-remembrance of some events considered of little or even null importance by the subject. These biases can lead a researcher to make erroneous associations and thus to reach erroneous conclusions [10, 22].

**Table 7.1** Characteristics of a possible case

| | YES | NO |
|---|---|---|
| **The patient shows clinical signs?**<br>The patient presents clinical manifestations typical of the event under study. | | |
| **Are the signals visible?**<br>The clinical manifestation of the event under study shows visible signs. | | |
| **Is the patient under medical observation?**<br>The patient is under medical observation caused by the event under study. | | |
| **Is there presumptive diagnosis?**<br>A presumptive diagnosis has been made based on the manifestations of the event. | | |
| **Did the doctor perform an exploratory examination?**<br>The doctor performed an exploratory examination to confirm the presumptive diagnosis. | | |
| **Was the patient asked to undergo a definitive test?**<br>The patient was asked to undergo a test in order to establish a definitive diagnosis of the event. | | |
| **Was the test performed?**<br>The definitive diagnostic test was performed to confirm the occurrence of the event. | | |
| **Was the result of the test positive?**<br>The definitive test was positive with respect to the occurrence of the event. | | |

In order to counteract the selection bias, Feinstein proposes that each case must consecutively fulfill a series of characteristics that each possible case must possess to be considered as such. Table 7.1 shows the series of characteristics [24].

## 7.7 Advantages and Disadvantages of Case–Control Studies

Cases-control studies are the appropriate choice and sometimes the only alternative to study diseases with very low incidence that affect the older adults, as in the case of systemic lupus erythematosus in men. This type of study design offers the possibility of detecting most of the cases of a particular disease in a specialty hospital and of subsequently investigating its possible causes, which, in the case of systemic lupus erythematosus, may be the consumption of certain medications [25, 26].

In a cohort study, it would be necessary to observe a very large number of older men to detect the few incidents of systemic lupus erythematosus in a given population, probably placing older adults at risk due to the use of medications, particularly those who consume ticlopidine [26], which could induce the development of systemic lupus erythematosus. This would make a cohort study very ineffective, since enormous resources and time would have to be invested in the follow-up of older men who are free of the disease; moreover, it would have unethical implications.

Case-control studies are also more useful than cohort studies in the case of diseases that, although not so rare, have a prolonged latency period, as in the case of bladder and kidney cancer and the study of its possible association with tea con-

**Table 7.2** Advantages and disadvantages of case-control studies

| |
|---|
| **Advantages** |
| They allow for the study of uncommon events and diseases, or those with a long latency period |
| They can be used when an estimate of prior risk is required to carry out a prospective study |
| They are faster to carry out |
| They have lower costs |
| They require fewer individuals |
| It is possible to use records as sources of information: Medical records, hospital discharges, morbidity statistics (potentially), specialized medical records (cancer and other chronic illnesses), records of epidemiological monitoring systems, death certificates or equivalent documents. |
| They pose no risk to individuals |
| They allow to study multiple potential causes of the event or disease |
| They can be used to assess population-based procedures or interventions (screening or immunization programs) |
| They can be used to study health conditions that require a relatively fast treatment |
| They constitute an alternative for when a prospective approach has unethical implications |
| **Disadvantages** |
| They depend on the memory of individuals or information records to determine past exposures (memory bias) |
| The validation of data could be difficult or even impossible |
| There is a higher chance of selection and information bias |
| They are not suitable for estimating the prevalence and incidence of an event or disease |
| The time lapse between exposure and disease cannot be easily established |
| They make it difficult to make a detailed study of the causal mechanisms of an event or disease |

sumption in older adults [27]. A case-control study is better in this case because the identified cases have already developed the disease and there is no need to wait for the period between exposure (tea consumption) and the manifestation of the disease (bladder and kidney cancer).

This type of study design makes it possible to study a wide range of exposures related to a given disease, such as the study of risk factors for fall-related fractures in older adults [6]. Another very important advantage is that it requires smaller samples and is less expensive than experimental designs or cohort studies [8].

Case-control studies can also be used in the assessment of population-based procedures or interventions, such as the evaluation of the clinical efficacy of a pneumococcal pneumonia vaccine in older adults [28].

One of the disadvantages or drawbacks of this type of study is that it can only provide information about the target event or disease in the population that has been selected to carry out the study.

In contrast with infectious processes, the causal agents of many chronic diseases are unknown, making it difficult to make a diagnosis and to distinguish between diseased and non-diseased subjects. The long latency typical of chronic diseases allows the involvement of many environmental and constitutional factors, making it difficult to determine with precision the moment of exposure to each of them [29].

Another factor that complicates the identification of cases (incidence) is the indefiniteness of the time at which clinical manifestations appear after exposure to the risk factor, which also varies from one individual to another.

This is important, since it is undeniable that there is a greater chance for bias and erroneous inference in this type of study, compared to other designs. The classification of exposures is subject to memory biases when making use of information recalled by the study subjects, which in this case are older adults [8, 9].

It is known that the loss of memory in older adults is caused by cognitive changes associated with normal aging [30], and this could introduce a greater memory bias when making use of information recalled by older adults. This bias could be reduced by seeking the help of a reliable informant (spouse, child, primary caregiver).

Another disadvantage of case-control studies is that they are not population-based, and thus the incidence and prevalence of the disease cannot be calculated directly [2]. Table 7.2 shows a summary of the advantages and disadvantages of case-control studies.

## 7.8 Conclusions

In the planning phase of studies related to aging and old age, the quality of knowledge derived from epidemiological studies depends, in an important way, on the reliability of the methodological design and the data collection strategies used to answer the research question. Case-control studies constitute a cost-effective alternative that allow to make valid and reasonably accurate estimates of the strength of the association between hypothetical causes and effects in studies of older adults.

## References


1. Bolúmar-Montrúll F (2011) Estudios de casos y controles. In: Hernández-Aguado I, Gil de Miguel A, Delgado-Rodríguez M (eds) Manual de epidemiología y salud pública : para grados en ciencias de la salud, 2nd edn. Médica Panamericana, Madrid, pp 135–145
2. Rothman KJ, Greenland S, Lash TL (2008) Modern epidemiology, 3rd edn. Wolters Kluwer Health/Lippincott Williams & Wilkins, Philadelphia
3. McGwin G Jr, Sims RV, Pulley L, Roseman JM (2000) Relations among chronic medical conditions, medications, and automobile crashes in the elderly: a population-based case-control study. Am J Epidemiol 152:424–431. https://doi.org/10.1093/aje/152.5.424
4. Mahillo-Fernandez I, de Pedro-Cuesta J, Bleda MJ, Cruz M, Mølbak K, Laursen H et al (2008) Surgery and risk of sporadic Creutzfeldt-Jakob disease in Denmark and Sweden: registry-based case-control studies. Neuroepidemiology 31:229–240. https://doi.org/10.1159/000163097
5. Yao S, Till C, Kristal AR, Goodman PJ, Hsing AW, Tangen CM et al (2011) Serum estrogen levels and prostate cancer risk in the prostate cancer prevention trial: a nested case-control study. Cancer Causes Control 22:1121–1131. https://doi.org/10.1007/s10552-011-9787-7

6. Coutinho ES, Fletcher A, Bloch KV, Rodrigues LC (2008) Risk factors for falls with severe fracture in elderly people living in a middle-income country: a case control study. BMC Geriatr 8:21. https://doi.org/10.1186/1471-2318-8-21
7. Christensen S, Mehnert F, Chapurlat RD, Baron JA, Sorensen HT (2011) Oral bisphosphonates and risk of ischemic stroke: a case-control study. Osteoporos Int 22:1773–1779. https://doi.org/10.1007/s00198-010-1395-y
8. Lazcano-Ponce E, Salazar-Martinez E, Hernandez-Avila M (2001) Case-control epidemiological studies: theoretical bases, variants and applications. Salud Publica Mex 43:135–150
9. González A, García-Rodríguez LA (2003) Estudios de cohortes y de casos y controles: qué podemos esperar de ellos. Gastroenterol Hepatol Contin 2:44–48
10. Gordis L (2014) Chapter 10 case-control and other study designs. In: Gordis L (ed) Epidemiology, 5th edn. Elsevier/Saunders, Philadelphia, pp 189–214
11. Tambyraja AL, Sengupta F, MacGregor AB, Bartolo DC, Fearon KC (2004) Patterns and clinical outcomes associated with routine intravenous sodium and fluid administration after colorectal resection. World J Surg 28:1046–1051; discussion 51–2.doi:https://doi.org/10.1007/s00268-004-7383-7
12. Bosco JL, Antonsen S, Sørensen HT, Pedersen L, Lash TL (2011) Metformin and incident breast cancer among diabetic women: a population-based case-control study in Denmark. Cancer Epidemiol Biomark Prev 20:101–111. https://doi.org/10.1158/1055-9965
13. Mittleman MA, Maclure M, Sherwood JB, Mulry RP, Tofler GH, Jacobs SC et al (1995) Triggering of acute myocardial infarction onset by episodes of anger. Determinants of Myocardial Infarction Onset Study Investigators. Circulation 92:1720–1725
14. Freedman DM, Dosemeci M, Alavanja MC (2000) Mortality from multiple sclerosis and exposure to residential and occupational solar radiation: a case-control study based on death certificates. Occup Environ Med 57:418–421. https://doi.org/10.1136/oem.57.6.418
15. Wingo PA, Higgins JE, Rubin GL, Zahniser SC (eds) (1994) An epidemiologic approach to reproductive health. WHO, Geneva
16. Elashoff JD, Lemeshow S (2005) Sample size determination in epidemiologic studies. In: Ahrens W, Pigeot I (eds) Handbook of epidemiology. Springer, Heidelberg, pp 559–594
17. Gómez-Gómez M, Danglot-Banck C, Huerta-Alvarado SG, García-de-la-Torre G (2003) El estudio de casos y controles: su diseño, análisis e interpretación, en investigación clínica. Rev Mex Ped 70:257–263
18. Baumgartner RN, Koehler KM, Gallagher D, Romero L, Heymsfield SB, Ross RR et al (1998) Epidemiology of sarcopenia among the elderly in New Mexico. Am J Epidemiol 147:755–763
19. Sánchez-García S, Juárez-Cedillo T, Gallegos-Carrillo K, Gallo JJ, Wagner FA, García-Peña C (2012) Frecuencia de los síntomas depresivos entre adultos mayores de la ciudad de México. Salud Ment 35:71–77
20. Lee LK, Shahar S, Rajab N, Yusoff NA, Jamal RA, Then SM (2013) The role of long chain omega-3 polyunsaturated fatty acids in reducing lipid peroxidation among elderly patients with mild cognitive impairment: a case-control study. J Nutr Biochem 24:803–808. https://doi.org/10.1016/j.jnutbio.2012.04.014
21. Suzuki T, Yoshida H, Hashimoto T, Yoshimura N, Fujiwara S, Fukunaga M et al (1997) Case-control study of risk factors for hip fractures in the Japanese elderly by a Mediterranean osteoporosis study (MEDOS) questionnaire. Bone 21:461–467
22. Dadonienė J, Žagminas K, Beržanskytė A (2013) Introduction to research methodology. Vilniaus universitetas: Vilniaus universiteto leidykla, Vilnius
23. Delgado-Rodríguez M, Llorca J (2004) Bias J Epidemiol Community Health 58(8):635–641
24. Feinstein AR (1979) Methodologic problems and standards in case-control research. In: Ibrahim MA (ed) The case-control study consensus and controversy, Pergamon, pp 35–41
25. Bosch X, Formiga F, López-Soto A (2012) Lupus eritematoso sistémico en el anciano. Rev Esp Geriatr Gerontol 47:71–75. https://doi.org/10.1016/j.regg.2011.11.005
26. Yokoyama T, Usui T, Kiyama K, Nakashima R, Yukawa N, Kawabata D et al (2010) Two cases of late-onset drug-induced lupus erythematosus caused by ticlopidine in elderly men. Mod Rheumatol 20:405–409. https://doi.org/10.1007/s10165-010-0289-3

27. Bianchi GD, Cerhan JR, Parker AS, Putnam SD, See WA, Lynch CF et al (2000) Tea consumption and risk of bladder and kidney cancers in a population-based case-control study. Am J Epidemiol 151:377–383
28. Vila-Corcoles A, Salsench E, Rodriguez-Blanco T, Ochoa-Gondar O, de Diego C, Valdivieso A et al (2009) Clinical effectiveness of 23-valent pneumococcal polysaccharide vaccine against pneumonia in middle-aged and older adults: a matched case-control study. Vaccine 27:1504–1510. https://doi.org/10.1016/j.vaccine.2009.01.013
29. Hernandez-Avila M, Garrido F, Salazar-Martinez E (2000) Biases in epidemiological studies. Salud Publica Mex 42:438–446
30. Budson AE, Price BH (2005) Memory dysfunction. N Engl J Med 352:692–699. https://doi.org/10.1056/NEJMra041071

# Chapter 8
# Longitudinal Studies and Older Adults Cohorts

**Carmen García-Peña, Claudia Espinel-Bermúdez, Pamela Tella-Vega, Mario Ulises Pérez-Zepeda, and Luis Miguel Gutiérrez-Robledo**

**Abstract** Science never before faced such a complex, dynamic and time-dependent process as human aging. Longitudinal studies are a source of fundamental evidence of multi-factor changes over time, especially those that have contributed to understanding the aging process through research questions related to the course or prognosis of physical or cognitive functioning of older adults, exposure to comorbidity, health conditions, and biological, environmental, social or emotional negative or positive factors, as well as other questions related to aging.

However, these studies have major methodological challenges to keep the validity of information between standardized measurements and the generalization of the results, especially with the loss of participants due different causes. These difficulties motivated the realization of this chapter where we discussed the role of the longitudinal studies on aging, starting with methodological concepts, the importance of this design in geriatric research and the direction of new research questions, we present also a review of classic longitudinal studies taken from literature, which enable us to provide examples of scope and methodological implications, finally we suggested some strategies about strengthen the validity and generalization of results.

Longitudinal methodology represents a fundamental pillar in geriatric research. Its implementation always must to be supported by good planning that takes into account-standardized procedures as well as techniques that minimize the probable losses during the follow-up having less effect throughout the study.

C. García-Peña (✉)
Research Division, National Institute of Geriatrics, Mexico City, Mexico
e-mail: mcgarcia@inger.gob.mx; mcgarciapena@gmail.com

C. Espinel-Bermúdez
Research Unit in Clinical Epidemiology, West Medical Center, Mexican Institute of Social Security, Guadalajara, Jalisco, Mexico
e-mail: mclaudia_espinel@yahoo.com.mx

P. Tella-Vega · M. U. Pérez-Zepeda
Department of Geriatric Epidemiology, National Institute of Geriatrics, Mexico City, Mexico
e-mail: ptella@inger.gob.mx; mperez@inger.gob.mx

L. M. Gutiérrez-Robledo
National Institute of Geriatrics, Mexico City, Mexico
e-mail: lmgutierrez@inger.gob.mx



C. García-Peña et al. (eds.), *Aging Research - Methodological Issues*,
https://doi.org/10.1007/978-3-319-95387-8_8

**Keywords** Longitudinal studies · Longitudinal-research · Follow-up studies

## 8.1 Introduction

The aging of the population is one of the most remarkable success stories in social development and human health. However, scientific understanding of the aging process has not developed at the same rate as the growth of the generations of older adults and their special health needs.

In this context longitudinal studies take on special importance, especially those that have contributed to understanding the dynamics of the aging process by analyzing physiological, psychological, social or environmental variables [1] that are time variant. These designs have yielded results on successful aging, longevity, frailty and other traits. Thus, geriatric research has been nourished by a diversity of studies of an observational and especially a longitudinal nature, oriented to responding to the lack of knowledge about the latent changes in different generations of older adults, and posing new questions about exceptionally healthy old populations linked to the traits of robustness and functioning, in contrast to populations that are frail or disabled/dependent, with or without chronic degenerative illnesses [2].

The present chapter will be concerned with discussing the role longitudinal studies play in the study of aging. First it will analyze the theoretical concept of the longitudinal study, and then it will highlight the importance of this design in geriatric research and the direction of new research questions. In the second part we present a review of classic longitudinal studies from the literature, which will enable us to provide examples of scope and methodological implications, in order to offer some strategies to strengthen the validity and generalizing of results.

## 8.2 Theoretical Concept of Longitudinal Studies

The broadest notion of longitudinal studies refers to the analysis of a particular sample of individuals who show time-dependent patterns of change (variables of interest), which require the presence of three conditions: (1) that the data be collected during two or more distinct time periods; (2) that the sample elements (individuals) are comparable from one period to the other; and (3) that the analysis involves comparison of the data between two or more time periods [3].

The chapter will be limited to showing general aspects of observational longitudinal studies, experimental studies like clinical trials or quasi-experimental studies are discussed in Chap. 9.

All epidemiological studies may be classified according to the way the information search will be oriented. Considerable debate around a unique epidemiological studies classification has been present in the medical literature. Pearce for example, discusses that all epidemiological studies are bases on a specific population and during a specific period of time. Consequently, author argues that the fundamental distinction is between studies of incidence and prevalence [4]. Truth is that there is not only one approach and we may to accept that more than one classification is needed depending on different purposes.

Related with longitudinal studies, also several perspectives had presented different classifications. We present below the most frequently accepted structure and characteristics of each one:

Panel studies [5] obtain information by repeated measurements of the same group of individuals over fixed periods of time. This kind of study represents the conceptual base of a census or national survey carried out within the same population, with certain time periodicity to answer questions about the change of latent variables through time. It is also used to distinguish permanent characteristics from transitory ones of a specific phenomenon, analyze the life conditions of a group being studied, or differentiate intergenerational changes that are presented in a stage of life. These could be functional dependency, retirement from the labor force or characteristics of longevity in a population, as well as others. Caruana et al. [6] separate panel studies in three types: i) cohort panels, where some or all individuals with similar exposures or impacts are considered over time, ii) representative panels, where data are collected for a random sample regularly and iii) linked panels, where data collected for other purposes is tapped and linked to form individual-specific datasets.

One of its main weaknesses is that the sample responses could be subordinated to a "period effect," caused by an unexpected event or general circumstance (epidemic, climate change, or civil unrest, for example) at the time of the measurements, which could change the responses issued differentially among the subjects of the study. As well, the panel study could have a significant decline in the number of responses in each cycle of information collection, losses which would have a cumulative effect on the study's variables. This is related to the progressive loss of members of the sample during the course of the study [5], a phenomenon common to any longitudinal study. These losses must be given special consideration in studies on aging, where it happens more frequently, since the losses are related to events such as address change, death, hospitalization of the participant, or a decision to stop participating. It causes the sample to get smaller in each measurement period, a phenomenon called "panel fraying" or attrition. Given the importance of this possibility it will be described in more detail at the end of the chapter.

Trend designs [5] differ from panel studies because they analyze changes through time of different individuals in each evaluation period. With this characteristic, the data collected are analyzed collectively and not individually. Thus, as the name indicates, the information analyzed enables researchers to predict future trends about the individuals or the study universe, and the prediction variables may be evaluated through time. In this type of study unforeseen factors in the sample subjects are not considered, but the results may be easily influenced by other time-related variables not considered. For this reason, this type of design must be clearly delimited and the information collection strategies must be strictly replicated in each measurement period.

Lastly, the cohort design [7] is the longitudinal study most used on epidemiology and clinical research, since it is thought to be closer to experimental studies in terms of the search for causality and scientific evidence [8]. However, for the social and demographic sciences the cohort study represents the measurement of differences or changes in a population (or group) selected according to a common condition or experience where the main point is rooted in analysis, together with the population and the magnitude of the time-dependent change of events, which is often defined by the researcher [3]. This definition has a common denominator with other scientific areas, in that the longitudinal data are compiled in a time sequence that clarifies the direction as well as the magnitude of change in the variables.

For epidemiological and clinical research, the cohort design makes it possible to check for a cause-effect association through time between the course of an exposure and an outcome of interest (event) that is produced over a period of time, where each subject makes an individual contribution [9, 10]. Like the other longitudinal studies, its objective is to describe the occurrence of time-dependent results [10], but the scope of this type of epidemiological study is related more to the incidence of phenomena from the composition of the groups for an exposure variable that is present or absent among the subjects being studied. It follows through time in a prospective or retrospective manner up to the appearance of the event of interest. Other cohort design studies worth to mention are case cohort design [11], integrates a selection of cases and a smaller set from an original cohort is and the nested case-control, within a cohort study, where cases and controls are selected from the cohort.

Retrospective cohort study or historical cohort occurs when the investigator accesses a historical roster of all exposed and non-exposed persons and then determines their current case/non-case status. The investigator initiates the study when the disease is already established in the cohort of individuals, usually long after the original measurement of exposure. In relative longitudinal studies on aging, there are few variables that might respond to complex questions related to time changes, and the exposure events of interest occurred in the past, in other words before the study was begun. That is why the researcher does not have control over the nature and quality control of prior measurements or over data that could be important for a specific question and that were not gathered in the past. For that reason, the study of diverse events in aging such as functioning, cognition, memory, depression, levels of physical activity, changes in body composition and others requires a prospective methodology that allows the researcher to include changes in the present time from their identification in repeated observations, for periods of time established by the researcher or until an outcome of interest is presented.

In a prospective cohort design investigator begins by integrating a sample of participants, measure some characteristics that can predict an outcome and then follows these participants and measure them periodically [12].

It must be taken into account that older persons differ from other groups of populations in several ways. With this in mind, we present a proposal outline of scenarios that could be represented in a longitudinal study within the scope of research on aging. This outline was prepared taking as a reference the proposal of Fuller for the

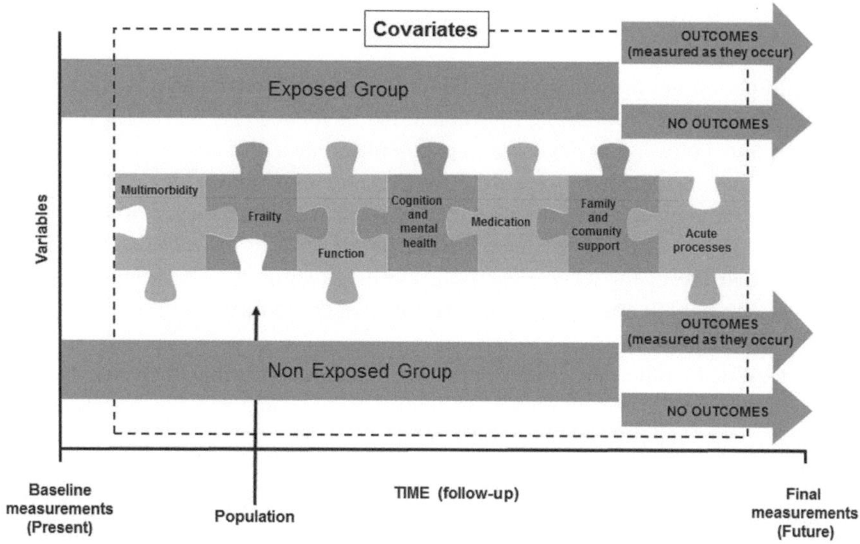


**Fig. 8.1** Proposed outline of the scenario of a longitudinal study on aging

study of falls among old people [13] and the analysis of levels of complexity present in geriatric research among different populations of older people by Faes et al. [1].

Figure 8.1 starts by identifying variables to be evaluated in the study, both in the population of interest or group being studied and in the comparison group. At the start of the study (present time) initial measurements are made that enable the researcher to define the state or presence of specific variables, to show later the change of the variables of interest or new outcomes differentiated between the group being studied and the comparison group. This makes it possible to identify causal relationships between individuals with or without a specific risk factor (cause), and to attribute the changes related to the outcomes (effect). Different covariates or potential confusing factors that could interrelate among the groups in the follow-up are evaluated in parallel, as well as the presence of normal changes in the process of aging that are present and interconnected throughout the study in such a way that, if the perspective of aging in these studies were not considered, they could establish spurious and non-causal relations between the initial state of the variables of interest, their exposure, and their final outcomes [14, 15] (Fig. 8.1).

A clear example of the theoretical concept of this design is presented by the Canadian Longitudinal Study on Aging (CLSA). It [16, 17] is considered one of the most important studies in present and future geriatrics because it has the possibility of following up a representative sample of the Canadian population over 20 years and will evaluate many medical and non-medical factors in the aging process (biological, psychological, social, lifestyle and economic). It will analyze how illness, health and well-being are influenced in older adults, and thereby achieve a better understanding of aging. The study began recruiting men and women between the ages of 45 and 85 years in 2010.

## 8.3 Advantages and Disadvantages of Longitudinal Studies

Longitudinal studies present important advantages over other observational designs (Table 8.1). It makes it possible to evaluate the incidence of a particular illness or outcome and helps when investigating potential causes of a possible outcome, whereby evidence is shown from the follow-up of exposed and non-exposed subjects at the moment they present an event being studied for the first time, or when it is modified by action of the exposure. In addition, it reduces bias between the exposure and the event when observing these sequentially. This type of design allows researchers to evaluate multiple results that could be related to the exposure factor [15].

Longitudinal studies with older population have helped with the understanding of the many complicated relationships among primary and secondary risk factors and health outcomes. Given that older adults have an increasing risk of adverse outcomes (of death and disability, for example) compared with other age groups, and numerous physio-pathological processes can be almost simultaneous, longitudinal observation of the facts during aging is of great value for scientific research [1, 2, 18, 19].

On the other hand, longitudinal studies share the disadvantages of observational studies. The interpretation of causal relationships can always be limited by the presence of many confusing variables. However, longitudinal studies offer the best observational option for studying causal inference. Other disadvantages are the lack of control in allocation of exposure (the realm of experimental studies), which could bring about differential biases in factors related to the occurrence of the event within the exposed and non-exposed group [9] and lastly, the follow-up brings with it uncontrollable losses for the researcher.

In studies based on the older population, the final statistical power of the sample is often affected for many health-related reasons. These could include problems like mortality rates of up to 20% per year if people older than 70 years of age are included, hospitalization and reports of illness, disablement and accidents, all of which commonly account for high rates of non-response and therefore losses for the

**Table 8.1** Advantages and disadvantages of longitudinal studies [6]

| Advantages | Disadvantages |
|---|---|
| Establishes cause-effect relationships in real time | No differential classification for lack of control in assigning of the exposure. Often requires large sample sizes |
| Reduces the presence of biases between exposure and the event | Selection bias through losses during the follow-up (morbidity or mortality of the seniors) |
| Evaluates measurements of incidence of an event, relative risk, excess risk. | Interdependence of time and variables related to the exposure and/or event |
| Observation of multiple results related to an exposure factor | Frequent use of key informants (proxies) |
| Efficient for unusual exposures | Complicated for infrequent events that are presented over a long period of time. Often very expensive |

study. As well, unlike what is likely to happen with other age groups, there may be causes of a social nature that could affect the quality of the information between measurements such as: changes of address, given that old people often move from one house to another since they depend on their support and care network, or frequent changes of a key informant with consequent difficulty in obtaining consistent and valid information.

The definition of "baseline evaluation" could also constitute a serious dilemma when dealing with old people. This is because of a complex interdependence with time and relationships among individual factors and measurements of results that could fluctuate even at time zero in a study, involving the exposure factor and the event being studied, along with other associations of change. These must be taken into account for an adequate interpretation of the results. Thus, when working with older populations the causal relationships are multiple and bi-directional, and the qualification of exposure must be identified with care, based on a solid theoretical framework.

The changes or modifications in the exposure factor or added variables that influence the exposure as much as the event of interest through time require broad and organized field logistics as well as more financial resources to maintain the cohort [9, 20]. Despite the advantage of being able to study several results, they could turn out to be difficult to analyze when one result produces a secondary one, and so on.

In some cases, the chain of causality is not clear or there is no conceptual consensus about certain topics, as is the case for issues like geriatric syndromes, frailty and the sequence between functioning and frailty. The inefficiency of longitudinal studies for studying "rare" or "low-frequency" events is an important point when discussing the senior citizen population. This means data collection requires extended periods of time during which the possibility increases of encountering correlated events that make it difficult to define the predictive factors. For that reason, it is fundamental to analyze clearly the type of topic that can be evaluated using this methodology, as will be discussed later.

## 8.4 The Importance of Longitudinal Studies in Research on Aging

It is a fact that aging and its relationship with demographic, social and technological changes have created important knowledge needs at every level, especially in the area of health. An example of this is the longitudinal study of several pathologies with aging that break the classic patterns of its natural history compared with younger generations [2]. Another is the study of variables that have a positive impact on the health of senior citizens for achieving longevity or successful aging or failing to do so, and the presence of events such as loss of functioning, cognitive deterioration and frailty. These studies help to meet the need for understanding aging from different perspectives [21–24].

With respect to the evaluation of interventions, although controlled clinical trials provide most of the evidence for evaluating their effectiveness, this type of design faces important ethical and logistical dilemmas, especially when evaluating older populations [24]. The presence of cognitive deterioration, multiple morbidity, polypharmacy, and other factors are variables that often limit the inclusion of these participants in clinical trials, in addition to the difficulty encountered when using a proxy to obtain informed consent [25, 26]. For those reasons observational follow-up of interventions added to the normal care of old people represents a viable alternative to carrying out clinical trials, where longitudinal evaluations of programs are especially important.

## 8.5 Methodological Scopes and Implications of Longitudinal Studies on Aging

Longitudinal studies have made important contributions to understanding aging [24], with research questions related to the course or prognosis of physical or cognitive functioning of old people, exposure to comorbidity, health conditions, and biological, environmental, social or emotional factors both negative and positive. One point that has also attracted much interest is the scientific history of large-scale longitudinal studies in which groups from young populations have been included who are evaluated right through until they reach old age, thus providing evidence of latent changes over time and the multiple relationships with their environment. Those contributions base their knowledge about the course of life in the older population not only on the demographic evolution of the cohorts but on biological changes, roles and socio-cultural needs presented through time and dependent on the different stages of development of human beings from birth to death [27, 28].

To analyze the changes caused by aging over time and the relevant issues for that stage of life, a systematic review that included 51 longitudinal studies of aging was taken as a reference point [28], identified from the data base of longitudinal studies of the United States federal government's National Institute on Aging (NIA). This enabled the researchers to establish six non-exclusive topics of frequent interest. In 44% of the studies questions were asked about cognitive function, 51% on health and physical performance, 55% on socio-economic status and 63% of the cases analyzed predictors of multi-morbidity and mortality. However, it is interesting that areas considered important such as health costs or genetic factors were not reported among the most frequent themes.

The authors of this review postulate that the guidelines of the longitudinal studies on aging should be broader to provide strategic information on health systems for the care for old people. Table 8.2 summarizes the topics identified by the authors as being those most frequently considered in the studies didactically included in their review. For this chapter issues are included in place of variables or measurement scales as expressed in Table 8.3 of the original article. Despite the fact that the topics

**Table 8.2** Topics related to longitudinal studies on aging [20]

| Topics identified | Related sub-themes |
|---|---|
| Cognitive function | Age and mortality, cognitive deterioration, mortality and quality of life, use of services and results in health, social roles |
| Socioeconomic status | Relationship between functional condition and morbidity-mortality analysis among self-perception of physical health, age, sex and conditions of life |
| Health and physical performance | Functional association performance and decline of health condition, disability as a predictor of mortality, gait speed, grip strength and balance. Index of body mass associated with coronary illness, falls, cognitive deterioration, hospitalizations and mortality |
| Predictors of morbidity-mortality | Relationship of the state of health and mortality, as well as markers of Inflammation (CRP, IL-6) as predictors of morbidity-mortality |
| Costs of health care | Individual effects relative to social networks, association between illness and dependence, health care of the old person and its impact on frailty and mortality |
| Epigenetics | Genetic causes of aging, status of health and genomic sequence associated with the state of health, APOE and risk of dementia |

CRP=C-reactive protein; IL-6 = interleukin 6

are analyzed relatively frequently, their relationship with other variables is reported less frequently. This is the case of the impact of cognitive decline in the results on health and the use of services, or the effect of social determinants on quality of life. As for clinical questions, what stood out was those biological variables that could become predictors of morbidity-mortality as metabolic, hormonal, immunological or other measurements that require time-dependent analysis.

To exemplify the approach and methodology of longitudinal studies in aging, we present below a review of some of the projects that we believe to be representative of this type of study [29–42]. We should mention that the examples presented correspond to follow-up studies more than to the classic epidemiological cohort design that include the definition of an exposure. A brief summary of these studies is presented in Table 8.3. The central themes are similar: identifying functional, social and environmental variables as predictors that change outcomes in aging. The average time of follow-up was 10 years.

In Mexico the study with the longest follow-up is the National Study of Health and Aging in Mexico (MHAS) [29] with four measurements over 14 years. The MHAS is a panel study representative of the Mexican population of subjects born in 1951. The purpose of this study is to evaluate the aging process of the Mexican population, especially changes in morbidity, disability, intergenerational transference systems, migration and economy, for which measurements were carried out. In 2001, 15,402 interviews were completed, directly or with a proxy, with a response rate of 93%. In 2003 the survey included 14,386 subjects, and in 2012 it included those interviewed in 2003 plus a new sample of persons born between 1952 and 1962, for a total of 18,465 persons, the fourth round was completed in 2015. Two more waves are prepared for 2018 and 2021.

**Table 8.3** Longitudinal studies on aging

| Study | Population | Follow-up | Objective |
|---|---|---|---|
| ENASEM-(MHAS in English) [29] Mexican study on health and aging in Mexico | n = 15,402 50 or more years of age | 2001 2003 2012 2015 | To obtain information on various characteristics of the objective population living in Mexico |
| ELSA [30] English longitudinal study of aging | n = 11,391 50 years of age or more | 2002-to present | Multidisciplinary approach related to health, well-being, financial and social resources, quality of life and deoxyribonucleic acid (DNA) to correlate the samples with epidemiological data |
| HRS [31] Health and retirement study | n= >37,000 50 years of age or more | 1992-to present | To understanding interaction between health and social, economic, and psychological circumstances, particularly with the retirement decisions. |
| PREHCO [32] The Puerto Rican elderly: Health conditions | n = 4, 291 60 years of age or more | 2002 2003 2006 2007 | To provide quality data for researchers and policy makers about issues affecting the elderly population in Puerto Rico |
| CRELES [33] Costa Rican Longevity and healthy aging study | n = 2820 Born in 1945 or before | 2005 2007 2009 | To determine the duration and quality of life, and its causal factors, of the Costa Rican elderly |
| ELSI [34] The Brazilian longitudinal study of ageing | n = 10,000 50 years of age or more | 2015-to present | To measure outcomes across a wide range of domains and to provide high-quality multidisciplinary data that can shed light on the causes and consequences of outcomes of interest. |
| SHARE [35] The survey of health, ageing and Retirement in Europe | n= >120,000 50 years of age or more | 2004-to present | Micro data on health, socio-economic status and social and family networks, covers 27 European countries and Israel. |
| TILDA [40] The Irish longitudinal study on ageing | n = 8504 50 years of age or more | 2010-to present | To provide and evidence base for addressing current and emerging concerns associated with aging population in Ireland |
| LASI [37] Longitudinal aging study in India | n = 60,250 45 years of age or more | 2012-to present | To provide comprehensive longitudinal evidence base on health, social and economic wellbeing of elderly population in India. |
| CHARLS [36] China health and retirement longitudinal study | n = 17,500 45 years of age or more | 2011-to present | To collect a high quality nationally representative sample of Chinese residents. |
| JSTAR [38] Japanese study of aging and retirement | n = 4200 Between 50 and 75 years of age | 2007 2009 2011 | Researchers can track the characteristics of the Japanese elderly population in terms of both their specificity and universality in the world. |

(continued)

**Table 8.3** (continued)

| Study | Population | Follow-up | Objective |
|---|---|---|---|
| KLoSA [41]Korean longitudinal study of aging | n= > 10,000 45 years or older | 2006 2008 2010 2012 2014 | To create the basic data needed to devise and implement effective social, economic policies to address the trends that emerge in the process of population ageing |
| IFLS [42] Indonesian family life survey | n= > 30,000 26 years of age or more | 2007 2008 | The fourth wave was redesigned to collect data similar to those in the HRS and similar surveys. |

The English Longitudinal Study of Aging (ELSA) [30] and the MHAS took representative samples from their country of origin. The average age of the cohorts when they entered the study was 50 years. As for methodological aspects, the ELSA included five measurements at two-year intervals, the first one in March 2002 with 11,391 subjects and their spouses (n = 708), chosen from the base of participants in the Health Survey of England (HES), a transversal survey carried out between 1998 and 2001. The criteria of eligibility were: having been born before March 1, 1952, participated in the HES, and lived in a private house at the time of the first measurement; the last measurement was done in 2011 with 10,317 subjects, with a response rate of 78%. The lack of response was minimized from subsequent imputations.

The Health and Retirement Study (HRS) [31] was conducted for 25 years. The HRS is a national representative survey of subjects over 50 years of age in the USA, provides evidence of psychosocial content and policy changes in addition to the expansion into biomarkers and genetics that may affect individuals. The initial HRS Cohort (1992), included individuals born between 1931 and 1941 and their spouses of any age. In order to make the sample fully representative of the USA some new cohorts were enrolled through the years, the last was added in 2010 (Mid baby boomers born 1954–1959). The main survey occurs every 2 years, making 2016 the 14th follow up.

The PREHCO Project [32] investigated the characteristics of older non-institutionalized adults over 60 years of age in Puerto Rico, cross-sectional sample survey of target individuals and their spouses. The first round was between 2002 and 2003 (2167 variables and 4291 cases). At this moment has completed a second round to become a longitudinal study. The follow-up questionnaire included questions regarding the changing conditions of those individuals who participated in the first round, this second wave has 4291 cases and 2766 variables, those participants deceased or institutionalized were interviewed using a proxy.

Another survey distinguished by extensive measurement of health indicators as well as biomarkers is CRELES [33], also has linkages with the Costa Rican National Death Index in the follow up mortality events. First wave (2005) include participants who born in 1945 or before (2827 participants), the second wave in 2007 revisited the same participant group and the third wave was conducted in 2009 with 1855 surviving participants. Through these rounds different measurements were made: fasting blood and overnight urine collection, DNA has been extracted for second and third cohort.

ELSI-Brazil [34] collects data relating to health and functioning, economic circumstances, social participation and networks, use of health services of adults aged 50 years or older. After the baseline information collection, the follow-up waves are biannual. This study has been planned to survey about 10,000 Brazilians every 2 years, new participants are added to refresh the different cohort age groups.

SHARE [35] is a multidisciplinary panel database of micro data on health, for more than 80,000 individuals in European population over 50 years of aged, this panel covers 27 European countries and Israel. From the first wave this database combined self-reports health with physical measurements reports. In wave four in 2010 were included dried blood spots samples, the relevance of this was that these blood spots were analyzed for C reactive protein, HbA1c and total cholesterol. In the future it will be possible to include vitamin D and inflammatory markers of the cytokine family.

TILDA [40] is a detailed study on ageing undertaken in Ireland. It involves detailed interviews of people aged 50 years and over, charting their health, social and economic circumstances over a 10-year period. As some of the previous international studies, TILDA also collected different important biomarkers.

LASI [37] is a study conducted in 30 states and 6 union territories of India, covering a panel sample size of 60,250 individuals aged 45 years and older and their spouses. This study adopts multi-stage clustering sampling design; three-stage sample design in rural areas and a four-stage sample design in urban areas. The main reason for adopting this four-stage sample is to make easier the selection of households because urban wards are quite large, making it difficult to list all the households.

CHARLS [36] is based on the HRS and related aging surveys such as the ELSA and SHARE. The pilot sample was a two-province sample collected in 2008 and followed up in 2012.The national wave of CHARLS was fielded in 2011and includes about 10,000 households and 17,500 individuals in 150 countries and 450 villages committees. These individuals will be followed up every 2 years.

JSTAR [38] is an interdisciplinary data resource on health, economic position, and quality of life in Japan. Like SHARE, HRS and ELSA this survey has the common aim to understand a variety of levels in aging, from individuals to countries, from an international perspective. The first wave of JSTAR took place in 2007, contains data on individual living circumstances of 4200 adults between 50 and 75 years of age. Three waves were conducted from 2007 to 2011.

In Korea: KLoSA [41] is sample of more than 10,000 persons at least 45 years of age that were interviewed among other topics, about information on work and income and health and disability. It was first conducted in 2006, with biennial follow-up waves since then. Topics include in this survey have an important impact on the economic and social activities of this group.

On the other hand, it is IFLS [42], a longitudinal survey representative of about 83 percent of the Indonesian population, containing more than 30,000 individuals in 13 in the country. First wave was conducted in 1993, it was designed as a household survey of individuals at least 26 years of age. For its fourth wave in 2007, it was

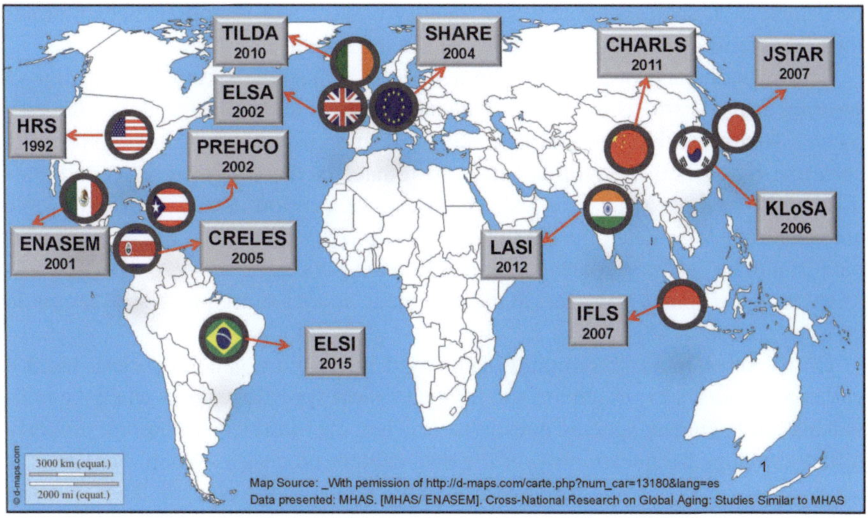


**Fig. 8.2** Longitudinal studies on aging [43, 44]

redesigned to collect data similar to those in the HRS and similar surveys, including community survey and local conditions. Fig. 8.2 shows geographical location and starting year of the previously mentioned longitudinal studies on aging [43].

## 8.6 Strategies for Improving the Validity and Generalization of Longitudinal Study Results

One of the objectives of the longitudinal studies is to recreate, from an initial exposure or measurement, the natural history or trend of an illness or event at the time it occurs. To do that, measurement stages are established that enable the identification of changes in a particular group that is followed through time [25]. Most phenomena related to aging are time-dependent [24, 45], as much in their appearance as in their duration, so it is necessary to predict the changes over time or the cumulative effect of multiple associations with respect to the intervals of the measurements done, for example variables associated with functional or cognitive decline, or predictors of morbidity-mortality or frailty.

Another peculiarity of longitudinal studies in geriatric population is the speed of change of age-related variables that might be presented in cutoff periods or jump between subsequent measurements. Thus, the basic threat to this design is centered on the study's losses from any number of causes. There is a lack of response in key variables [46, 47] before the presence of unexpected adverse events that could arise in the development of the longitudinal studies, such as death, hospitalization, disability of a participant or even changes in geographic mobility, for which the data are lacking, and so losses in longitudinal studies become a frequent challenge. For this reason, prior to the start of the study, strategies for containment of losses should be established, such as strategies of quality control and retention that minimize these problems and ensure the validity of the information.

This phenomenon of losses during the course of a study is known as attrition or wearing away. It affects the sample size and makes it difficult to calculate the estimators, such as making an adequate statistical inference. As well, it could result in selection bias when the participants who remain in the study present conditions different from those who were lost. The causes of attrition in studies of old people are often related to death, hospitalization or disability that occur when the participant cannot be evaluated during the measurement period. However, the participants might reappear in a third or fourth measurement, so that subsequent analyses could be more complex [48].

On the other hand, the missing data could be due to a general pattern when a participant refuses to participate. Or, it might follow specific patterns when the individuals fail to answer specific questions, or when the interviewer does not properly follow the steps for questioning, or the data capture process is wrong. For that it is necessary to analyze some variables that make it possible to contrast them with the sample in general and could be estimated if the absence of data affects the internal validity of the study [47, 48].

There are strategies that could be planned and executed during fieldwork, such as home visits, telephone contact, or the incorporation of interviews with proxy informants such as primary care givers or the participant's spouse, who could provide responses close to what would have been given by the participant. Fieldwork could be enriched by using retention methods (positive messages, contact on important dates, information on the progress of the study, etc.) with the goal of minimizing the fatigue of staying in a follow-up over many years. A special element when working with older populations is offering the profile of interviewers, which demonstrates interest and empathy with the interviewees. It is thought to be useful in retaining subjects if the same interviewers are present during the different measurements, since that could create a climate of confidence with older adults and avoid rejection in subsequent measurements.

With the real possibility of failure to get information or sustaining losses during follow-up (whether they are occasional or constant), analytic strategies are required that reliably estimate the measurements made. However, it must be kept in mind how important it is to plan a study of this magnitude properly, especially in key sections that will be tied to the occurrence of losses, as well as including supervision strategies of data quality that enable later analysis to minimize the losses.

### 8.6.1 Sampling and Sample Size

It is crucial to make the right decision on how the sample in the first measurement will be set up, since errors committed in this phase will be very difficult to correct later. In this sense it must be ensured that each sample unit or individual is chosen randomly from the sample framework as a probability sample, to increase the precision of the study by being able to ensure that the samples are really independent right from the start of the study. As well, the sample size must in all cases take into account a percentage of real losses consistent with the theme of the study or area of influence of the participants. It is necessary to have a large enough number of observations, given that attrition tends to reduce the number of individuals over time, and so a size that will allow for the occurrence of said losses must be ensured. That way, the analysis may be carried out in the sub-groups of the population of interest without exposing the statistical power of the study at risk.

### *8.6.2 Standardization of the Measurements*

As discussed earlier, longitudinal studies are based on the change of variables through time, for which the presence of random errors in the measurements represents a very complex problem, which could even overestimate the final results of the measurements. In this situation is especially useful to include not only previously validated and standardized instruments, but also to carry out exhaustive training of the field personnel, adding control questions that permit the analysis of differences at a given time and reduction of false data of change in the variables of interest, and carrying out a periodic calibration of the measurement equipment.

### *8.6.3 Imputing Lost Data*

This requires the implementation of generalized equations, for which there are several statistical methods. In general, it is thought that in the first instance variables should be created that identify the data "without response or without measurement" in each measurement cut-off. This dichotomy of variables will serve as sub-groups within the study for the key variables, the reason for the imputation [47]. These variables must be compared between the defined times as losses (for example, an initial or baseline measurement vs. the second or third vs. the fourth measurement). This type of analysis considers the measurements between periods to be dependent, since they are from the same subjects, as happens in the case of the panel type study or the design of cohort study, so they must be seen as paired statistical tests, in order to establish whether or not there are statistically significant variations between the measurement times. If not, it will verify that the losses did not affect the behavior of the variables; otherwise, if it proves that the data do have variations between measurements, a multivariate probit model will be integrated that predicts the probability of attrition conditional on a set of variables measured in each cut-off during the study. This model identifies the common source of the data variation and is integrated as a possible response to the matrix model in seeking the most common responses. In any case it can mathematically predict the variability of the error, and if this does not have statistical significance, the data are presented as being free from error [46, 49, 50].

### 8.6.4 Data Weighting

This process is fundamental for an adequate estimate of the data. Reasons for the need to weigh the variables involved include the lack of responses, unequal selection of groups, adjustments in medication, and others.

Weighting involves giving each sample unit a numeric value that would be representative of its population being studied. Thus, the weight of each variable in particular includes the relative value of the sub-sample it represents and the relationship between the size of the sample and the proportion of subjects interviewed. A process of statistical inference for each variable or time period involved is developed from these values. It should be mentioned that weighting in a longitudinal study could include cross-sectional weighting (within the same measurement time) or between subsequent measurements time zero vs. n times involved [31, 51].

### 8.6.5 Harmonization of Data

A very important effort was launched on 2015 by the RAND Corporation and the National Institute on Aging (NIA) which convened a meeting of the Network on the Harmonization of International Aging Studies with support from to facilitate cross-national comparisons [52]. The web page "Gateway to Global Aging Data" (www.g2aging.org) aims to promote harmonization and to serve as a repository of information about the HRS-family of studies. This web page presents comparative descriptions of the HRS-family of studies, harmonized variables and datasets, among other tools. Several longitudinal studies around the world are now present in this webpage.

## 8.7 Conclusions

Science had never before faced such a complex, dynamic and time-dependent process as human aging. Longitudinal studies are a source of fundamental evidence of the multi-factor changes over time, which enables it to maintain the evaluation of interventions that have a timely and positive impact on the course of aging in the population.

Longitudinal methodology represents a milestone in geriatric research. Its implementation always must be backed up by good planning that takes into account standardized procedures as well as techniques that minimize the probable losses during the follow-up and the consequent effect throughout the study. Finally, it will be expected that the results derived from the follow-up will reflect the evidence of a phenomenon present in the senior citizen population.

## References


1. Faes M, Van Iersel M, Olde Rikkert M (2007) Methodological issues in geriatric research. J Nutr Health Aging 11(3):254–259
2. Tappen RM, Ouslander JG (2010) State-of-the-art in longitudinal studies on aging: an overview of the supplement. J Am Geriatr Soc 58(Suppl 2):S283–S286. https://doi.org/10.1111/j.1532-5415.2010.02912.x
3. Ruspini E (1999) Longitudinal research and the analysis of social change. Qual Quant 33:219–227
4. Pearce N (2012) Classification of epidemiological study designs. Int J Epidemiol 41(2):393–397. https://doi.org/10.1093/ije/dys049
5. Trivellato U (1999) Issues in the design and analysis of panel studies: A cursory review. Qual Quant 33(3):339–351
6. Caruana EJ, Roman M, Hernández-Sánchez J, Solli P (2015) Longitudinal studies. J Thorac Dis 7(11):E537–E540. https://doi.org/10.3978/j.issn.2072-1439.2015.10.63
7. Eldredge J (2002) Cohort studies in health sciences librarianship. J Med Libr Assoc 90(4):380–392
8. Furst DE (2004) Observational cohort studies and well controlled clinical trials–we need them both! J Rheumatol 31(8):1476–1477
9. Burch TK (2001) Longitudinal research in social science: Some theoretical challenges. Can Stud Popul 28(2):263–283. https://doi.org/10.25336/P6H30P
10. Hernández-Avila M, Garrido-Latorre F, López-Moreno S (2000) Diseño de estudios epidemiológicos. Salud Publica Mex 42(2):144–154
11. The Pennsylvania State University (n.d) Advanced Cohort Study Design [Internet]. Available from: https://onlinecourses.science.psu.edu/stat507/node/62
12. Hulley SB, Cummings SR, Browner WS, Grady D, Newman TB (2013) Designing clinical research, 4th edn. Lippincott Williams and Wilkins, Philadelphia
13. Fuller GF (2000) Falls in the elderly. Am Fam Physician 1 61(7):2159–2168 2173–4
14. McNutt L-A, Wu C, Xue X, Hafner JP (2003) Estimating the relative risk in cohort studies and clinical trials of common outcomes. Am J Epidemiol 15 157(10):940–943
15. Lazcano-Ponce E, Fernández E, Salazar-Martínez E, Hernández-Avila M (2000) Estudios de cohorte. Metodología, sesgos y aplicación. Salud Publica Mex 42(3):230–241
16. Kirkland SA, Griffith LE, Menec V, Wister A, Payette H, Wolfson C et al (2015) Mining a unique Canadian resource: The Canadian longitudinal study on aging. Can J Aging 24 34(3):366–377. https://doi.org/10.1017/S071498081500029X
17. Ma J, Thabane L, Beyene J, Raina P (2016) Power analysis for population-based longitudinal studies investigating gene-environment interactions in chronic diseases: A simulation study. PLoS One 22 11(2):e0149940. https://doi.org/10.1371/journal.pone.0149940
18. RAND Corporation (n.d) Center for the Study of Aging [Internet]. RAND Labor and Population Available from: https://www.rand.org/labor/aging.html
19. National Center for Health Statistics (n.d) Longitudinal Studies of Aging [Internet]. Available from: https://www.cdc.gov/nchs/lsoa/index.htm
20. Healy P, Devane D (2011) Methodological considerations in cohort study designs. Nurse Res15 18(3):32–36. https://doi.org/10.7748/nr2011.04.18.3.32.c8461
21. Bowling A (2002) Research methods in health : Investigating health and health services, 2nd edn. Open University Press, Philadelphia 486 p

22. Carlson MDA, Morrison RS (2009) Study design, precision, and validity in observational studies. J Palliat Med 12(1):77–82. https://doi.org/10.1089/jpm.2008.9690
23. Schaie KW, Hofer SM (2001) Longitudinal studies in aging research. In: Birren JE, Schaie KW (eds) Handbook of the psychology of aging, 5th edn. Academic Press, San Diego, pp 53–77
24. Newman AB (2010) An overview of the design, implementation, and analyses of longitudinal studies on aging. J Am Geriatr Soc 58(Suppl 2):S287–S291. https://doi.org/10.1111/j.1532-5415.2010.02916.x
25. Rochon PA, Gurwitz JH, Sykora K, Mamdani M, Streiner DL, Garfinkel S et al (2005) Reader's guide to critical appraisal of cohort studies: 1. Role and design. BMJ 6 330(7496):895–897
26. Kaufman SR, Shim JK, Russ AJ (2004) Revisiting the biomedicalization of aging: clinical trends and ethical challenges. Gerontologist 44(6):731–738
27. Walter R, Heinz VWM (eds) (2003) Social dynamics of the life course: transitions, institutions, and interrelations, 1st edn. Transaction Publishers, New York 306 pp
28. Stanziano DC, Whitehurst M, Graham P, Roos BA (2010) A review of selected longitudinal studies on aging: Past findings and future directions. J Am Geriatr Soc 58(Suppl 2):S292–S297. https://doi.org/10.1111/j.1532-5415.2010.02936.x
29. Mexican Health and Aging Study (MHAS) (2015) Methodological Document, Mexican Health and Aging Study [Internet] 2015. Available from: www.MHASweb.org
30. Steptoe A, Breeze E, Banks JNJ (2013) Cohort profile: The English longitudinal study of ageing. Int J Epidemiol 42(6):1640–1648. https://doi.org/10.1093/ije/dys168
31. Sonnega A, Faul JD, Ofstedal MB, Langa KM, Phillips JWR, Weir DR (2014) Cohort profile: The Health and Retirement Study (HRS). Int J Epidemiol 43(2):576–585. https://doi.org/10.1093/ije/dyu067
32. Dávila AL, Garcia A, Larriuz M, Reyes L, Palloni P, McEniry M. La salud de los adultos de edad mayor en Puerto Rico. Informe General 2002–2003. [Internet]. Puerto Rico; 2004. (Informe General Proyecto PREHCO.). Available from: http://prehco.rcm.upr.edu/sites/default/files/website_pdf/Inform1.pdf
33. Rosero-Bixby, Luis, Xinia Fernández, William H Dow. CRELES: Costa Rican Longevity and Healthy Aging Study, 2005. ICPSR26681-v1; Report No: 2010–07–21
34. Lima-Costa MF, de Andrade FB, de Souza PRB, Neri AL, de Oliveira Duarte YA, Castro-Costa E et al (2018) The Brazilian longitudinal study of aging (ELSI-BRAZIL): objectives and design. Am J Epidemiol. https://doi.org/10.1093/aje/kwx387/
35. Alcser KH, Benson G, Börsch-Supan A, Brugiavini A, Christelis D, Croda E et al (2005) In: Börsch-Supan A (Coord), Jürges H (eds) The survey of health, ageing and retirement in Europe – methodology, 1st edn. Munich, Mannheim Research Institute for the Economics of Aging 355 p
36. Chen X., Smith J., Strauss J., Wang Y. ZY. China health and retirement longitudinal study. In: Pachana N., eds. Encyclopedia of geropsychology. Living Ed. Singapore: Springer Singapore; 2015. p. 1–8
37. Arokiasamy P, Bloom D, Lee J, Feeney K, Ozolins M (2012) Longitudinal aging study in India: vision, design, implementation, and some early results. In: Smith JP, Malay Majmundar E (eds) Aging in Asia: findings from new and emerging data Initiatives, 1st edn. The National Academies Press, pp 36–74
38. Ichimura H, Hashimoto H, Shimizutani S. Japanese study of aging and retirement: JSTAR first results 2009 [Internet]. Tokio; (2009). Available from: https://www.rieti.go.jp/jp/publications/dp/09e047.pdf
39. Kenny RA, Whelan BJ, Cronin H, Kamiya Y, Kearney P, O'Regan C et al (2010) In: Barrett A, Finucane C, Timonen V (eds) The design of the Irish longitudinal study on ageing. Dublin, Trinity College Dublin 149 pp
40. Cronin H, O'Regan C, Finucane C, Kearney P, Kenny RA (2013) Health and aging: development of the Irish longitudinal study on ageing health assessment. J Am Geriatr Soc 61(Suppl 2):269–278. https://doi.org/10.1111/jgs.12197

41. Jang SN (2016) Korean Longitudinal Study of Ageing (KLoSA): overview of research design and contents. In: Pachana N (ed) Encyclopedia of geropsychology. Springer Singapore, Singapore, pp 1–9
42. Strauss J, F Witoelar, B Sikoki, AM Wattie (2009) The Fourth Wave of the Indonesia Family Life Survey: Overview and Field Report. Indonesia. 92 p. (WR-675/1-NIA/NICHD)
43. MHAS/ENASEM.(n.d) Cross-National Research on Global Aging: Studies Similar to MHAS
44. http:/d-maps.com/carte.php?num_car=13180&lang=es
45. Harman D (1981) The aging process. Proc Natl Acad Sci 78(11):7124–7128
46. Palmer RF, Royall DR (2010) Missing data? Plan on it! J Am Geriatr Soc 58(Suppl 2):S343–S348. https://doi.org/10.1111/j.1532-5415.2010.03053.x
47. Ayala YMO (2007) Estimación de datos faltantes en medidas repetidas con respuesta binaria. Revista Colombiana de Estadística 30(2):265–285
48. Feng D, Cong ZSM (2012) Missing data and attrition. In: Newsom J, Jones RN, Hofer SM (eds) Longitudinal data analysis : a practical guide for researchers in aging, health, and social sciences. Taylor & Francis, London, pp 71–97
49. Erten-Lyons D, Sherbakov LO, Piccinin AM, Hofer SM, Dodge HH, Quinn JF et al (2012) Review of selected databases of longitudinal aging studies. Alzheimers Dement 8(6):584–589. https://doi.org/10.1016/j.jalz.2011.09.232
50. Delgado Rodríguez M, Llorca Díaz J (2004) Longitudinal studies: concepts and particularities. Rev Esp Salud Publica 78(2):141–148
51. AIHW, Logie H, Hogan R, Peut A (2004) In: AGE 42 (ed) Longitudinal studies of ageing: implications for future studies, 1st edn. AIHW, Canberra 217 p
52. Samuel Thomas, Rose Li, Associates Inc. RAND HRS Around-the-World Harmonization Meeting [Internet]. Maryland; (2015). Available from: http://www.rand.org/content/dam/rand/www/external/labor/aging/pdfs/RAND-Harmonization-Summary-2015.pdf

# Chapter 9
# Clinical Trials on Aging Research

**Mario Ulises Pérez-Zepeda, Antonio Cherubini, Carmen García-Peña, Elisa Zengarini, and Luis Miguel Gutiérrez-Robledo**

**Abstract** Clinical trials are considered to be one of the best methodologies in health research and they are used primarily to test interventions in medicine. Aging research is no exception for this goal, and clinical trials are used to test different interventions in older adults with a number of variations in this particular research. In addition to drugs, in older adult's diverse non-pharmacological interventions are experimented for a wide-array of diseases and conditions that are particular for this age group. A careful design and sometimes adaptation of clinical trials methodology are necessary to have accurate results and translate them into actions in everyday clinical care of the older adult. Below, we provide a general and schematic review of the theoretical concept of clinical trials and their variants, followed by examples of interventions and specific outcomes in research on older adults.

**Keywords** Clinical trials · Pragmatic trials · Interventions for older adults

M. U. Pérez-Zepeda (✉)
Department of Geriatric Epidemiology, National Institute of Geriatrics, Mexico City, Mexico
e-mail: mperez@inger.gob.mx

A. Cherubini · E. Zengarini
Geriatria, Accettazione geriatrica e Centro di ricerca per l'invecchiamento. IRCCS-INRCA, Ancona, Italy
e-mail: A.CHERUBINI@inrca.it; el.zangarini@hotmail.it

C. García-Peña
Research Division, National Institute of Geriatrics, Mexico City, Mexico
e-mail: mcgarcia@inger.gob.mx

L. M. Gutiérrez-Robledo
National Institute of Geriatrics, Mexico City, Mexico
e-mail: lmgutierrez@inger.gob.mx



C. García-Peña et al. (eds.), *Aging Research - Methodological Issues*,
https://doi.org/10.1007/978-3-319-95387-8_9

## 9.1 Introduction

Up to half of all drugs consumed by human beings are used by older adults, yet they are still underrepresented in clinical trials [1, 2] –the main source of scientific evidence that a substance can be safe and effective [3, 4]. Nevertheless, drug trials are not the only therapeutic interventions amenable to be tested in clinical trials, a number of non-pharmacological strategies (e.g. technology, nutrition, physical activity, etc.) are increasingly used in the clinical armamentarium for older adult health care. Recognizing how older adults differ from the rest of the population and need modified clinical trials will not only increase the availability of interventions but will also reduce the number of individuals that present to health services with adverse drug reactions –a rather burdensome problem in economic terms for the system [5]; or lack of an appropriate intervention (i.e. either pharmacological or non-pharmacological) [6].

Notwithstanding, many interventions used in treatment of older people have been mainly empirical or derived from existing evidence from trials with younger adults. Only in recent years has the understanding of the problems and effective interventions in this age group started to be enriched through an increasing proportion of clinical trials that involve older adults, developing health care of senior citizens toward geriatrics based on scientific evidence. In addition, the range or variations that clinical trials acquire within the group of seniors enriches the possible approaches that could be made in experimental design that responds to a question being researched [7]. As well as its use in the clinic (mainly through geriatrics), other interventions could also be evaluated with these types of designs, and their application extends from the management of technology to improve the functioning, support groups, health services, and even complete systems [8].

In large part, "conventional" clinical trials in younger adults are often used to test a drug for the cure or control of a specific illness. Among older adults the probability of suffering from "only" one illness is low, i.e. multimorbidity is common, so they are often systematically excluded from pharmacological trials. This is opposite to what should happen, considering that they are the main consumers of pharmacological products, and therefore of the data generated from these clinical trials [9, 10]. New strategies for closing this gap should be sought in order to increase the quality of care of older adults. On the other hand, this type of design is used for evaluation of other non- pharmacology therapeutic options, such as exercise, nutritional interventions, support groups, technology (gerontechnology), education, and more [11–13].

Moreover, studies of older adults that evaluate multicomponent (i.e., interventions with more than one therapeutic component, not necessarily interacting between each other) or complex interventions (i.e., with multiple therapeutic components that synergize between each other) are common, in contrast to conventional pharmacological clinical trials. In addition, it is important to use variations of the classical clinical trial design (pragmatic trials) in this age group to evaluate existing health services (Chap. 12 deals entirely with health services research in older adults) and cost-effectiveness studies [7, 14]. Furthermore, we need to mention studies where interventions directed to other than the individual afflicted with the target condition (e.g., caregivers, service providers, etc.) are carried, and impact the older adult health, even if these interventions are not applied directly to the sick individual [15].

Finally, having evidence-based treatments (either pharmacologic or non-pharmacologic), will improve the use of available interventions, diminishing the indiscriminate use of ineffective or marginally effective therapies that place older adults in a greater risk of having a deterioration of their health. It is important to mention that this chapter will only deal with individual clinical trials, for information on studies for health care systems, a detailed description can be found in the health systems chapter (Chap. 12).

## 9.2 Concept of Clinical Trials

Randomized clinical trials have been defined as "prospective studies that involve humans, evaluate the effectiveness of an experimental intervention when compared with a control group or standard intervention, or with two or more existing interventions; generally designed by recruiting several 100 persons and yielding evidence that could be sufficient to bring about a broad change in public health policies or in current standards of health care" [16]. Clinical trials are used for understanding the effects of interventions in human beings, such that a well-planned and managed clinical trial could be solid evidence for daily use in medicine [17]; in other words, show causality of the tested intervention [18].

In its simplest form, an intervention may be tested by noting the subsequent effect on the same group of persons, a design known as before-and-after. Usually before-and-after studies are used when an intervention is implemented; for example, in delirium prevention, and this implies that different individuals are evaluated in the before and in the after group. The main weakness is when the controls are historical, while the design is more robust when it is possible to evaluate the same group of subjects before implementing the intervention and then after. Moreover, this design has little validity due to various biases that could give a false impression of a change (i.e., regression to the mean, placebo effect, etc.). At the next level, is the application of a non-randomized intervention to two distinct groups (i.e., the control group and the intervened group). The main problem with this design is selection bias, since there is the option of choosing who will be given the new treatment. The classic scheme of a clinical trial features randomized subjects for assignment of the intervention with different variants as to control of the intervention, blinding the administration of the intervention and the sequence of administration of the intervention. This last characteristic is highlighted in the so-called cross-over trials, in which all the subjects are administered the experimental intervention. One group is the study's control group at the start of the study, and in a second round it receives the intervention (in many cases between the two periods there is something known as the washout period, to eliminate the residual effect of those who were submitted to the experimental intervention), and the outcomes at each of the times are measured. It is important to plan and design appropriately clinical trials for older adults, both taking into account particular issues of this group as well as standard items depicted in the latest SPIRIT report [18].

It is thought that a clinical trial has more internal validity if the assigning of the intervention is randomized (the results generated are precise), the intervention is controlled throughout the process and the blinding is at the maximum. However, the external validity (generalization of the results at the population level) with such a rigid design could be compromised. To resolve this type of methodological problem clinical trials have been classified in two large groups: explanatory (internal validity) and pragmatic (external validity) [14]. In research on older adults, the use of pragmatic trials makes it possible to test interventions that are not possible with clinical trial; moreover, it may be possible to evaluate larger population groups (see Table 9.1). In addition, pragmatic trials allow to assess interventions that are currently ongoing or common practice in clinical settings.

### *9.2.1 Explanatory Studies*

This is most commonly used in the pharmaceutical industry sponsored trials, with great internal validity but with critical difficulties at the time of large-scale implementation in population groups. Their intention is to ensure that the subject has the intervention constantly and requires high levels of adherence to the treatment and comparison with a standard or a placebo. The assigning of a treatment is one of the fundamental points, as well as blinding of most of the participants in the study (i.e., the ill older adult, caregivers, researchers, etc.). Trials of this kind are more difficult to carry out in this age group, given that their characteristics, i.e. multimorbidity, polypharmacy, geriatric syndromes and the high prevalence of non-adherence, and therefore not useful for generalized conclusions [19]. Although it has been used mainly for the evaluation of drugs, it can also be used in other types of interventions, such as nutrition, education, physical activity and others.

### *9.2.2 Pragmatic Studies*

Among these studies are those with greater external validity and that imitate "real" conditions in which a therapy is implemented. Unlike explanatory studies, they are often most useful in evaluating interventions outside of the pharmacological field, such as complex interventions (multiple components that interact with each other), socio-medical, rehabilitation, physical activity, nutrition, studies, etc. [20]. The evaluation of health care services is one of the main uses of this type of trial, since for a particular system that is already established, it promotes the evaluation of what could be achieved without having complete and strict control over the intervention

**Table 9.1** Differences between explanatory trials and pragmatic trials

| | Explanatory trial | Pragmatic trial |
|---|---|---|
| Recruitment | Direct recruitment, in general does not consider the service provider | If randomizing is used for groups, the patients and the service providers are used. This involves having consent and baseline data from both |
| Selection criteria | Strict inclusion and exclusion criteria to lower the probability of confusing factors that have an impact on the outcome (like illnesses) | Try to be as inclusive as possible to improve the possibility of generalizing the results |
| Randomization | Randomizing by participant | Randomizing by groups is often necessary, especially for replicating real clinical scenarios |
| Blinding | Double blind is considered obligatory | Blinding must be done wherever possible, but the nature of the intervention generally doesn't permit it. Blinding the surveyors, analysts and baseline data collection before randomizing could lower the bias for lack of blinding |
| Intervention | Frequently a single simple activity (taking a medicine) | Frequently involves a complex activity of interaction between patient and doctor. The doctors need only a few light instructions to carry out the intervention. |
| Control | Usually a placebo or a gold standard treatment | Usually standard care |
| Contamination | Avoided with blinding | The randomization of groups can lower its occurrence but it does not avoid it |
| Evaluation of adherence | High levels of adherence are required | Efforts to achieve adherence to the intervention are made; however, it should not exceed what would normally be done in the clinic. On the other hand, the adherence is analyzed as an outcome and could indicate that an intervention is not useful on a day-to-day basis |
| Sample size | Standard statistical calculation | The effects of randomization in groups should be taken into account for the calculation. Consequently, a higher number of subjects is required to reach sufficient power |
| Outcomes | Frequently specific | Also, very specific. Some secondary outcomes are used to try to explain the effect of the intervention. |
| Analysis | Univariant and multivariant analysis are standard | Take into account in the randomized analysis of the group, since the outcomes of a group could be similar for the medical effect |

but rather with an evaluation of its process and the determination of its components to be able to learn the potential measurers of effectiveness (or not) of the same on a determined outcome [21]. Further detail on these studies could be found on Chap. 12, where pragmatic trials are identified as 'implementation research'.

## 9.3 Recruitment and Drop-Out

This is one of the best studied phases of clinical trials on older adults, and also one of the more problematic. It has been shown that failure in recruitment ranges often between 10–50%, compared to a lower failure of recruitment in younger adults, i.e. 5–10% [22]. As stated previously, there is an ongoing trouble on having a good representation of older adults in clinical trials. Issues for an adequate recruitment arise both from *a priori* discrimination of older adults, but also from subjects themselves [23]. Some barriers and potential solutions are depicted in Table 9.2. Some authors have suggested to run a 'prospective preference assessment' when planning a clinical trial for older adults, in order to anticipate those that would more probably would participate in a given trial [24], this in addition to profiling and knowledge of which conditions will facilitate participation or increase the interest on participating, some examples are: flexible process, confidence created with the researcher or the research institution, allocate enough budget for this phase of the trial, involving of the family or the community in the trial, taking into account that older adults have also things to do (i.e. not taking for granted that they are in complete inactivity and waiting to participate in our trials), identify 'hard to reach' groups and design specific strategies to recruit, increase the knowledge of the social benefit of participating in clinical trials, etcetera [25].

**Table 9.2** Barriers and their potential solution for the participation of older people in pharmacological clinical trials (Adapted from Lindley et al. 2012)

| Barrier | Solution |
|---|---|
| Potential risk of commercial use | More public financing by the government and other institutions of pharmacological clinical trials in senior citizens |
| Exclusion of subjects with comorbidity | Include them but measure |
| | Do specific studies for old people with specific pathologies (cancer, frailty) |
| Limitation of access to the research site | Include home visits or transportation to the research center in the logistics |
| Family opposition to entering into a study | Invest resources and time in information sessions |
| High mortality rate | This could be an advantage in some cases, where any intervention could lower the rate. It could be a disadvantage for the high rate of losses attributed to mortality; it is necessary to make an estimate of an appropriate sample taking losses by mortality into account |

## 9.4 Interventions

As stated previously, older adults have a wide array of interventions that are used for ameliorating both complications of chronic disease and age-related conditions. Along with the classical pharmacological interventions, in which older adults are under-represented; a number of other non-pharmacological interventions will be described.

### *9.4.1 Pharmacological*

The use of medications in this group of population is well-known, where some people routinely consume more than five medications, so called polypharmacy, and the average consume more than two. This, unfortunately, is generated with evidence from other age groups (see Fig. 9.1).

As mentioned earlier, there are several ways to categorize the groups of older adults (see Chap. 1). However, the generation of new drugs must define the function of that division; in other words, medication for the chronically ill, for the disabled, and for those who are frail. There is a need also to determine which group of older adults would benefit the most of a determined intervention, for example, those older adults with a lower life expectancy would not benefit from a pharmacological treatment that will take years to benefit the individual (e.g., statins).

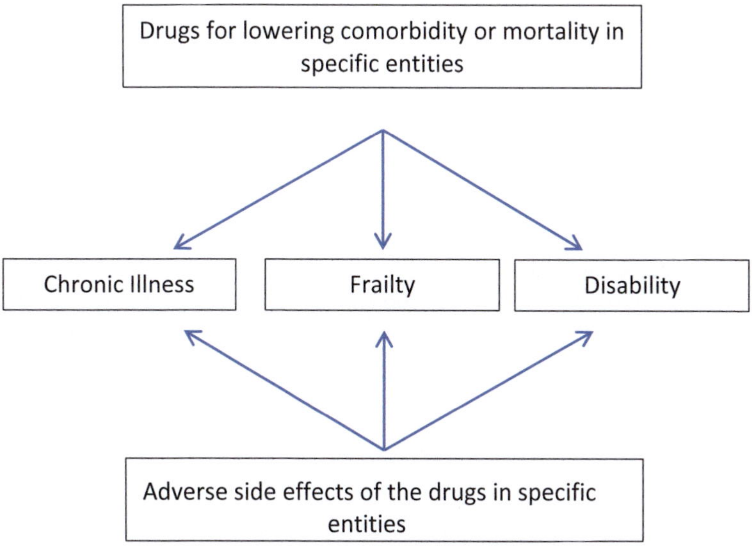


**Fig. 9.1** Spectrum of pharmaceutical clinical research in old people

There is evidence of exclusion of the older group from the clinical trials of the pharmaceutical industry [26] and in publicly funded trials. An example of the lack of correspondence between existing need and the creation of products for this need is the proportion of older subjects with cancer, which in countries like the United States is as high as 63%, while clinical trials have a representation of just 25%. Table 9.2 shows some of these barriers and their potential solutions [27]. In a recent review of clinical trials for senior citizens, Lindley calls for the inclusion of frail subjects in all pharmacological clinical trials; however, this always represents a risk, given the nature of frailty, where a priori a high level of side effects can be expected. However, this same review suggests always doing the measurement of frailty, so that the condition of the subjects included in the study can be known [19] (see Table 9.2); this should be added to already well-established characterization of older adults in clinical trials such as creatinine clearance or expected survival. In this respect the European Medicine Agency has recently released a reflection paper on frailty, to stimulate the measurement of this condition as a way to more appropriately identify vulnerability than chronological age (http://www.ema.europa.eu/docs/en_GB/document_library/Scientific_guideline/2018/02/WC500244285.pdf).

### *9.4.2 Deprescription*

Given the high frequency of side effects in old people, this type of trial is designed to stop a drug in some subjects and compare them with subjects that continue to use it: a drug previously identified as potentially inappropriate is withdrawn and its use is compared "inversely" with subjects who continue with the treatment, in many studies continuing to give a placebo to the group that has left the drug in which it hopes for a better outcome. This has been tested in subjects with dementia, especially with the antipsychotics, which have seen a higher frequency of mortality due to cerebrovascular events in large population studies.

An example of this type of study is called DART AD (*Dementia Antipsychotic Withdrawal Trial)*, in which patients with Alzheimer's are randomized into two groups, one that continues using the antipsychotic medication, and in the other group it is replaced with a placebo. A reduction was found in the mortality of the placebo group compared with those who continued with the antipsychotic medication [28, 29].

### *9.4.3 New Uses of Old Drugs*

Innovating tools are allowing to use old drugs with new purposes. Such is the case of metformin, currently used as an 'anti-aging' molecule. A number of questions arise in order to perform trials of drugs that have already passed thru a complete process and have been deemed safe. However, it is still not clear how this off-label or new uses could imply [30]. This 're-vamping' of older drugs has been used as an example of intervening to stop deleterious effects of aging, a central concept to what nowadays is called 'geroscience' (see Chap. 4). In addition, there are other interventions that are starting to be used, such as transfusions and the use of stem cells. The problem is that we do not have appropriate definitions and results of these trials, could be blur and do not provide a solution for older adult's ailments.

### 9.4.4 *Non-pharmacological Interventions*

These interventions include nutritional, physical activity, technology among others. Many trials evaluating these interventions have significant methodological problems. Even when solid evidence is available, these interventions are often not well known by health care professionals, thus limiting their adoption [31–34]. It is hard to describe each one, however they should always be taken into account due to its value in improving health in older adults. For example, a recent study on the use protein, showed that older adults do not benefit from higher than recommended protein intake in order to improve function and muscle mass. Generally speaking, they are more time consuming, and require expertise and control over the intervention fidelity.

### 9.4.5 *Complex Interventions*

Complex interventions have a series of processes that cannot be totally controlled, and that often are not pharmacological, but could include a component of pharmacological intervention. One classic study of these characteristics is from the Beswick group, where an intervention with many components was evaluated to learn its effect on the mobility in a group of ambulatory old people [20].

Other recent examples on complex interventions are the LIFE and the SPRINTT, both trials used a number of interventions tested to ameliorate the impact of frailty and sarcopenia, mainly based in nutritional support and physical activity. The first study has started showing results and the latter is in its last stages [35, 36].

## 9.5 Outcomes

In this category there isn't much debate on which outcomes should be used as effect variables in clinical trials in other population groups. However, in the case of old people, mortality, one of the outcomes most broadly used for other groups, isn't very useful with the advanced age due to the increase in the probability of death, a phenomenon that often presents a plateau around 90 years of age, where mortality starts to be equalized, and that could be observed graphically on the Gompertz curve. Among the main outcomes to test the effectiveness of an intervention on them, as well as mortality, are found to be functioning, dependence, geriatric syndromes (falls, incontinence, delirium, etc.), quality of life and utilization of services [8, 37].

## 9.6 Regulation

In addition to the well-known guidelines to conduct clinical trials (e.g. SPIRIT), and because of the particularities of the older adult, specific guidelines have been developed to properly develop medicines for geriatric patients. A number of guidelines are available in order to develop specific drugs for the older adults, however, currently there is a lack of regulation on how drugs should be tailored for this group of age. The example of needing something like this, is that regulation that is present for pediatrics [38]. This is of particular importance in order to have not only appropriate molecules with therapeutic value, but also the drug as a product that can be easily used. On the other hand, a recurring issue in the literature about clinical trials is the higher costs of researching for older adults, derived from all these care that is asked to properly develop trials. Therefore, a correct regulation on this matter could give more information to funding agencies on how trials for elderly are performed (please see Chap. 18 for a complete description on funding for aging research) and give enough money to perform them adequately [3].

For example, there is a recent interest in experimenting in older adults with interventions without solid evidence, such is the case of stem cells or transfusion for entities such as frailty or sarcopenia. Of special concerns of some of these trials is the requirement to pay a fee to enter an uncertain trial; appropriate regulation, certainly will halt this kind of experimentation [39].

Finally, regulatory issues seem to have impact in the decision of an older adult to participate in a clinical trial, as demonstrated by a recent study where older adult believing that the study was approved and reviewed by authorities was very important, were up to 3 times eager to participate in an influenza vaccine trial [40].

## 9.7 Conclusions

Clinical trials in research on older adults are enriched by distinct possibilities that exist for evaluating an intervention. What is of special interest is to learn these options for carrying out clinical research in this age group, so that they can have an extended reference of what could be obtained by utilizing this design on this population group.

Regulatory agencies all over the world should start recognizing older adults as the larger group of therapies consumer (either pharmacologic or non-pharmacologic), and therefor start the application of rules for clinical trials where available or elaborate legislation where there is still lack of it.

It is important to take into account the rights of older people in clinical trials, from the PREDICT consortium [6, 38]:

1. Older people have the right to access evidence-based treatments.
2. Promoting the inclusion of older people in clinical trials and preventing discrimination.
3. Clinical trials should be made as practicable as possible for older people.
4. The safety of clinical trials in older people.
5. Outcome measures should be relevant for older people.
6. The values of older people participating in clinical trials should be respected.

## References


1. Cherubini A, Del Signore S, Ouslander J, Semla T, Michel JP (2010) Fighting against age discrimination in clinical trials. J Am Geriatr Soc 58:1791–1796. https://doi.org/10.1111/j.1532-5415.2010.03032.x
2. Crome P, Lally F, Cherubini A, Oristrell J, Beswick AD, Clarfield AM, Hertogh C et al (2011) Exclusion of older people from clinical trials: professional views from nine European countries participating in the Predict study. Drugs Aging 28:667–677. https://doi.org/10.2165/11591990-000000000-00000
3. De Spiegeleer B, Wynendaele E, Bracke N et al (2016) Regulatory development of geriatric medicines: to GIP or not to GIP? Ageing Res Rev 27:23–36
4. He Z, Langford A (2017) Comparative analysis of geriatric and adult drug clinical trials on ClinicalTrials.gov. Stud Health Technol Inform 245:1265
5. Leporini C, De Sarro G, Russo E (2014) Adherence to therapy and adverse drug reactions: is there a link? Expert Opin Drug Saf 13(Suppl 1):S41–S55
6. Cherubini A, Cerenzia A, Zengarini E (2012) The exclusion of older patients from clinical trials regarding heart failure. Causes and consequences. Recenti Prog Med 103:103–108
7. Craig P, Dieppe P, Macintyre S et al (2008) Developing and evaluating complex interventions: the new medical research council guidance. BMJ 337:a1655
8. Dartigues JF (2005) Methodological problems in clinical and epidemiological research on ageing. Rev Epidemiol Sante Publique 53:243–249
9. Cesari M, Perez-Zepeda MU, Marzetti E (2017) Frailty and multimorbidity: different ways of thinking about geriatrics. J Am Med Dir Assoc 18:361–364
10. Cherubini A, Oristrell J, Pla X et al (2011) The persistent exclusion of older patients from ongoing clinical trials regarding heart failure. Arch Intern Med 171:550–556
11. Chernoff R (2005) Dietary management for older subjects with obesity. Clin Geriatr Med 21:725–733 vi
12. Mittelman MS, Ferris SH, Steinberg G et al (1993) An intervention that delays institutionalization of Alzheimer's disease patients: treatment of spouse-caregivers. Gerontologist 33:730–740
13. Theou O, Stathokostas L, Roland KP et al (2011) The effectiveness of exercise interventions for the management of frailty: a systematic review. J Aging Res 2011:569194
14. Patsopoulos NA (2011) A pragmatic view on pragmatic trials. Dialogues Clin Neurosci 13:217–224
15. Martin-Carrasco M, Martin MF, Valero CP et al (2009) Effectiveness of a psychoeducational intervention program in the reduction of caregiver burden in Alzheimer's disease patients' caregivers. Int J Geriatr Psychiatry 24:489–499

16. Freedman LS, Simon R, Foulkes MA et al (1995) Inclusion of women and minorities in clinical trials and the NIH Revitalization Act of 1993–the perspective of NIH clinical trialists. Control Clin Trials 16:277–285 discussion 286–279, 293–309
17. Friedman LM, Furberg C, Demets DL (2010) Fundamentals of clinical trials. Springer, New York
18. Chan AW, Tetzlaff JM, Altman DG et al (2013) SPIRIT 2013 statement: defining standard protocol items for clinical trials. Ann Intern Med 158:200–207
19. Lindley RI (2012) Drug trials for older people. J Gerontol A Biol Sci Med Sci 67:152–157
20. Beswick AD, Rees K, Dieppe P et al (2008) Complex interventions to improve physical function and maintain independent living in elderly people: a systematic review and meta-analysis. Lancet 371:725–735
21. Perez-Zepeda MU, Gutierez-Robledo LM, Sanchez-Garcia S et al (2012) Comparison of a geriatric unit with a general ward in Mexican elders. Arch Gerontol Geriatr 54:e370–e375
22. Ridda I, Macintyre CR, Lindley RI et al (2010) Difficulties in recruiting older people in clinical trials: an examination of barriers and solutions. Vaccine 28:901–906
23. Cherubini A, Gasperini B (2017) How to increase the participation of older subjects in research: good practices and more evidence are needed! Age Ageing 46(6):878–881. https://doi.org/10.1093/ageing/afx123
24. Kerman HM, Deshpande BR, Selzer F et al (2018) Willingness of older adults to participate in a randomized trial of conservative therapies for knee pain: a prospective preference assessment. Contemp Clin Trials Commun 9:93–97
25. Liljas AEM, Walters K, Jovicic A et al (2017) Strategies to improve engagement of 'hard to reach' older people in research on health promotion: a systematic review. BMC Public Health 17:349
26. Siu LL (2007) Clinical trials in the elderly–a concept comes of age. N Engl J Med 356:1575–1576
27. Ridda I, Lindley R, Macintyre RC (2008) The challenges of clinical trials in the exclusion zone: the case of the frail elderly. Australas J Ageing 27:61–66
28. Ballard C, Hanney ML, Theodoulou M et al (2009) The dementia antipsychotic withdrawal trial (DART-AD): long-term follow-up of a randomised placebo-controlled trial. Lancet Neurol 8:151–157
29. Ballard C, Lana MM, Theodoulou M et al (2008) A randomised, blinded, placebo-controlled trial in dementia patients continuing or stopping neuroleptics (the DART-AD trial). PLoS Med 5:e76
30. Podhorecka M, Ibanez B, Dmoszynska A (2017) Metformin - its potential anti-cancer and anti-aging effects. Postepy Hig Med Dosw (Online) 71:170–175
31. Abraha I, Rimland JM, Trotta FM, Dell'Aquila G, Cruz-Jentoft A, Petrovic M, Gudmundsson A et al (2016) Effectiveness of non-pharmacological interventions to prevent falls in older people: a systematic overview. The SENATOR Project ONTOP series. PLoS ONE 11(8):e0161579. https://doi.org/10.1371/journal.pone.0161579
32. Abraha I, Trotta F, Rimland JM, Cruz-Jentoft A, Lozano-Montoya I, Soiza RL, Pierini V et al (2015) Efficacy of non-pharmacological interventions to prevent and treat delirium in older patients: a systematic overview. The SENATOR project ONTOP Series. PLoS ONE 10(6):e0123090. https://doi.org/10.1371/journal.pone.0123090
33. Vélez-Díaz-Pallarés M, Lozano-Montoya I, Abraha I, Cherubini A, Soiza RL, O'Mahony D, Montero-Errasquín B et al (2015) Nonpharmacologic interventions to heal pressure ulcers in older patients: an overview of systematic reviews (The SENATOR-ONTOP Series). J Am Med Dir Assoc 16(6):448–469. https://doi.org/10.1016/j.jamda.2015.01.083
34. Abraha I, Cruz-Jentoft A, Soiza RL, O'Mahony D, Cherubini A (2015) Evidence of and recommendations for non-pharmacological interventions for common geriatric conditions: the SENATOR-ONTOP systematic review protocol. BMJ Open 5(1):e007488. https://doi.org/10.1136/bmjopen-2014-007488

35. Espeland MA, Gill TM, Guralnik J et al (2007) Designing clinical trials of interventions for mobility disability: results from the lifestyle interventions and independence for elders pilot (LIFE-P) trial. J Gerontol A Biol Sci Med Sci 62:1237–1243
36. Landi F, Cesari M, Calvani R, Cherubini A, Di Bari M, Bejuit R, Mshid J et al (2017) The "sarcopenia and physical fRailty IN older people: multi-componenT treatment strategies" (SPRINTT) randomized controlled trial: design and methods. Aging Clin Exp Res 29(1):89–100
37. Cherubini A, Bernabei R, Ferrucci L, Marchionni N, Studenski S, Vellas B (eds) (2015) Clinical trials in older adults. Wiley, New York, pp 1–280
38. Crome P, Cherubini A, Oristrell J (2014) The PREDICT (increasing the participation of the elderly in clinical trials) study: the charter and beyond. Expert Rev Clin Pharmacol 7:457–468
39. Robbins R (2018) How a society gala was used to sell young-blood transfusions to baby boomers desperate to cheat death. In: STAT. STAT, www.statnews.com, p Electronic Magazine Article
40. Akmatov MK, Jentsch L, Riese P et al (2017) Motivations for (non)participation in population-based health studies among the elderly - comparison of participants and nonparticipants of a prospective study on influenza vaccination. BMC Med Res Methodol 17:18

# Chapter 10
# Mixed Methods in Geriatrics and Gerontology Research

**Joseph J. Gallo and Jin Hui Joo**

**Abstract** Mixed methods research is defined as the collection, analysis, and integration of both quantitative data (e.g., outcome of a randomized trial, such as decrease in depression score) and qualitative data (e.g., observations or interviews about the experience of persons with depression) to provide a more comprehensive understanding of a research problem than might be obtained through quantitative or qualitative approaches alone. Because mixed methods research designs place high value on the stories behind the numbers – both in exploratory designs where the experiences and insights of the community under study inform the quantitative investigation, and in explanatory designs where they illuminate the quantitative data – mixed methods are especially appropriate in studying the problems of older adults and caregivers. The goal of this chapter is to introduce the concept of mixed methods as an approach for research in Geriatrics and Gerontology. We will discuss (1) why an investigator might consider using mixed methods; (2) basic and complex mixed methods designs; (3) examples of mixed methods projects, drawing upon our own research; and, (4) challenges in carrying out mixed methods research.

**Keywords** Mixed methods research · Complex mixed methods designs · Novel designs for aging research

J. J. Gallo (✉)
Mixed Methods Research Training Program for the Health Sciences, Department of Mental Health, Johns Hopkins Bloomberg School of Public Health, Baltimore, MD, USA
e-mail: jgallo2@jhu.edu

J. H. Joo
Department of Psychiatry and Behavioral Sciences, Johns Hopkins University School of Medicine, Baltimore, MD, USA
e-mail: jjoo1@jhmi.edu



C. García-Peña et al. (eds.), *Aging Research - Methodological Issues*,
https://doi.org/10.1007/978-3-319-95387-8_10

## 10.1 Introduction

Mixed methods research is defined as the collection, analysis, and integration of both quantitative data (e.g., outcome of a randomized trial, such as decrease in depression score) and qualitative data (e.g., observations or interviews about the experience of persons with depression) to provide a more comprehensive understanding of a research problem than might be obtained through quantitative or qualitative approaches alone [1]. Typical applications of mixed methods in the health sciences involve adding qualitative interviews to follow up on the outcomes of intervention trials, gathering both quantitative and qualitative data to assess patient reactions to a program implemented in a community health setting, or using qualitative data to explain the mechanism of a study correlating behavioral and social factors to specific health outcomes [2]. The goal of this chapter is to introduce the concept of mixed methods as an approach for research in Geriatrics and Gerontology. We will discuss (1) why an investigator might consider using mixed methods; (2) basic and complex mixed methods designs; (3) examples of mixed methods projects, drawing upon our own research; and, (4) challenges in carrying out mixed methods research.

## 10.2 What are Mixed Methods?

The term "mixed methods" refers to the integration of both quantitative and qualitative approaches into a single research project, drawing strengths from each approach. Quantitative methods, such as surveys, are able to estimate the prevalence of conditions and test the strength of associations among variables (see Chap. 5 for Descriptive Studies). Quantitative methods are derived from positivist assumptions about establishing cause and effect relationships (determinism), identifying key variables to describe a phenomenon (reductionism), measuring a construct (measurement), or testing a hypothesis (deductive logic) -- an *etic* or 'culturally neutral' or 'professional' perspective (see Chap. 7 and Chap. 8). As discussed in Chap. 6 qualitative approaches are framed from constructivist or realist worldviews that seek insight and interpretation of context at individual, social, and organizational levels – an *emic* or 'culturally unique' perspective [3]. Often quantitative approaches (exemplified by the fields of biostatistics and epidemiology) seek to generalize replicable results from the sample to a population (e.g., using a standardized depression questionnaire to estimate the prevalence of depression in the population from results in a sample, or describing characteristics of persons who decide to undergo a medical procedure). In contrast, qualitative approaches (exemplified by the fields of anthropology and sociology) seek to understand the scope of a domain (e.g., understanding the experience and concept of depression from the point of view of an individual from a certain culture, or why someone made a certain choice). Incorporation of quantitative and qualitative methods in a single research project may provide a more complete picture of phenomena or events [4].

The use of mixed methods in the health sciences is an international trend [5–8]. Mixed methods have found application in fields such as nursing [9], medicine [10], mental health [11], cardiovascular health [12], palliative care [13], public health [14], intervention development [15–17], implementation science [18, 19], health policy [20], global health [21], health disparities [22, 23], and gerontology [24]. Researchers developing interventions in these fields have found value in mixed methods approaches to evaluate responses to the intervention by stakeholders in the settings in which the intervention is to be delivered [25]. Mixed methods data can augment a randomized clinical trial or intervention design by gathering data before, during, or after a trial [2] to improve the development of an intervention, or to explain the outcomes of a trial [13]. Mixed methods approach operates as core methodologies of implementation science across the translational continuum [26–28].

## 10.3 Why Use Mixed Methods?

Health sciences research can benefit from a mixed methods approach in which qualitative and quantitative approaches are combined to enhance the study of health problems and strategies [29–31]. Borkan states that mixing qualitative and quantitative methods in a single study combines the benefits of generalizability with the thick, contextual interpretation of experience [31]. A limitation of much work is that it stops at the mixing of methods rather than considering different epistemologies [32]. In general, researchers have used qualitative methods to add more depth to previously defined constructs. However, "qualitative tokenism" [33] will not get to the core of what we need to understand about heterogeneity in patient expectations, values, and preferences that drive acceptance of treatment, adherence, or how we might modify services so that the most people get the most good – a public health model in contrast to a strictly medical model. Adding focus groups does not a mixed methods study make. Our experience has shown that an alternative to "qualitative tokenism" is to consider whether qualitative methods can help us understand individuals' experiences and the meanings they attribute to them without trying to "fit" them into previously defined constructs [34]. By allowing individuals to describe their own reality, we learn how people experience and define their distress differently than researchers, or how an intervention is adapted to specific contexts. Differing perspectives across patient cultural groups, families, clinician experiences and interests that act to affect how people respond to attempts to improve care demand that we consider new ways to carry out services research. Often investigators are only trained and comfortable with a single methodology, so the approaches to problems and the types of questions asked are constrained by training limited to one "world view" [35]. Reliance on a single methodological stance is no longer tenable in an increasingly complex multicultural and interdisciplinary context, or in the translation and dissemination of population and behavioral research to broader applications and conditions.

## 10.4 Basic Mixed Methods Designs

Purposeful and planned integration of quantitative and qualitative approaches is a key feature of mixed methods. Driven by the needs and goals of the research, investigators who consider specific and planned efforts to integrate quantitative and qualitative methods at all stages - study design, data collection, analysis, and interpretation – have the best chance to take full advantage of the promise of mixed methods. Figure 10.1 outlines some ways that qualitative and quantitative methods have been combined from a design perspective, and uses the terminology of Creswell and Plano Clark [36]. We want to emphasize at the outset that the depictions in Fig. 10.1 are only a few possible combinations, and investigators will want to consider carefully what points of integration and designs are best suited to the research question at hand.

In an **explanatory sequential design** (top of Fig. 10.1), quantitative data collection (and often data analysis) occurs before qualitative data collection. Investigators may use this design to help understand or explain quantitative findings. A statistical analysis of outcomes of a trial can tell what happened, but not why or how. An example of the use of an explanatory sequential design would be interviewing persons who did not benefit from the intervention (based on the quantitative assessments typically used in a trial) to find out how the intervention could be modified to be more successful. In an **exploratory sequential design**, qualitative data collection precedes quantitative data collection (second from the top in Fig. 10.1). This design may be familiar to many investigators as it is commonly used as a strategy to design an instrument or questionnaire or to gather information to guide the content of an intervention. The qualitative component helps "discover" or "uncover" the range of domains and the words people use to express ideas; participants are able to express attitudes, beliefs, feeling and constructs that are most important to them. The analysis of the qualitative data informs the next quantitative phase of the research.

In **concurrent designs** (middle of Fig. 10.1), less emphasis is placed on how one strand informs the next, and more on making inference or drawing conclusions from the concurrent strands. In a design in which the quantitative methods are primary, participants might undergo extensive structured assessments using standard assessment instruments (e.g., behavioral change measures) but some participants might be selected for more detailed evaluation (e.g., what a participant who received an intervention experienced), to understand processes in intervention implementation (e.g., what practitioners really do), or to study mediation (e.g., through the use of selected case studies to understand causal pathways [37]. Some mixed methods designs are sometimes described as an **embedded design** in which one component is emphasized and a much smaller component is incorporated into the data collection (bottom of Fig. 10.1).

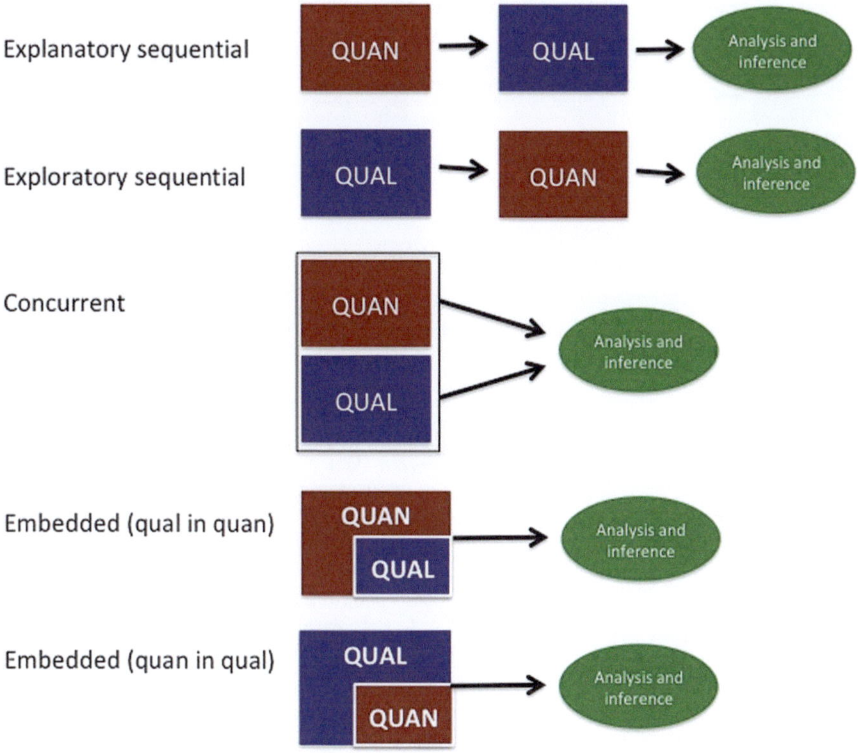


**Fig. 10.1** Basic mixed methods design

## 10.5 Complex Mixed Methods Designs

In this section we provide an introduction to complex mixed methods designs of the type used in developing and evaluating interventions. Careful consideration of Fig. 10.2 shows how the basic design ideas presented in the previous section can be combined to achieve specific study objectives. Experienced investigators described the added value of qualitative research in trials, explaining how it solved problems at the pretrial stage, explained findings, and helped to increase the utility of the evidence generated by the trial [38]. The complex designs presented here are by no means exhaustive, and the reader should conclude that mixed methods designs can be tailored to answer study aims in many different ways.

In the designs shown in Fig. 10.2, an exploratory sequential phase (qualitative followed by quantitative) is followed by an explanatory sequential phase (quantitative followed by qualitative). In **sequential mixed methods randomized trial design** (top of Fig. 10.2), an exploratory sequential stage to develop and refine the intervention is followed by an explanatory sequential stage to evaluate the intervention. In the first phase, the intervention is developed and adapted based on stake-

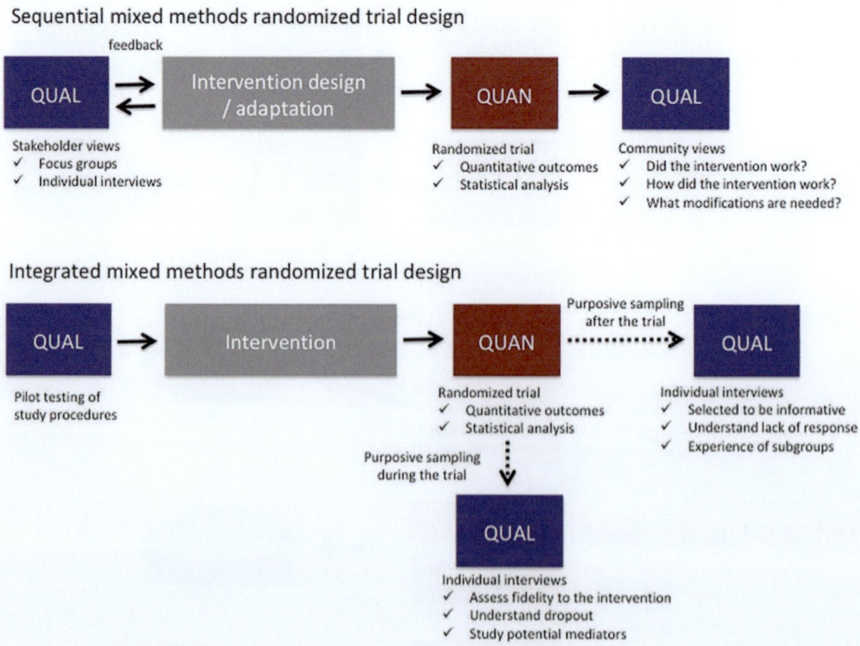


**Fig. 10.2** Complex mixed methods design

holder input derived from focus groups and individual interviews. For example, to make sure an intervention intended for older adults, investigators might elicit views of older adults in focus groups and carry out interviews with health care providers. In the second phase, a randomized trial may be carried out to study the effect of the intervention on outcomes such as depression scores or functional measures. Data from the trial is analyzed with statistical methods. In the qualitative phase (the far-right box at the top of Fig. 10.2), the community is again engaged to provide views about how the intervention worked and how the acceptability could be improved. For example, investigators might hold focus groups or community forums to present results of the trial and hear from community members how to improve the intervention for dissemination to new settings.

The bottom of Fig. 10.2 illustrates an **integrated mixed method randomized trial design**. We call this design "integrated" to emphasize that the sample for the qualitative phase at the end is integrated with the sampling of participants for the trial. In one case purposive sampling occurs after the trial is done (the far-right box at the bottom of Fig. 10.2). A "formative" or qualitative phase before the randomized trial goes into the field can be useful to pilot test and, if necessary, modify, recruitment, consent, and study procedures. Then during or after the trial, a qualitative phase can shed light on questions about how and why the intervention worked or fell short of expectations, using purposive sampling. A "purposive" sample is one that is sampled to be informative of a research question, not to be representative of all participants. For example, participants may be selected for open-ended qualitative interviews who did not respond to the intervention as expected, or to gain insight from subgroups about how the intervention worked (e.g., interviewing minority older adults). Purposive sampling could occur during the trial (box at the bottom of Fig. 10.2). For example, investigators might talk to participants who drop out to determine why, to assess how and why the intervention was carried out with fidelity, or to ask participants about factors in the intervention that seemed to be particularly helpful or unhelpful (i.e., potential mediators of the intervention effects).

## 10.6 Integration of Quantitative and Qualitative Analyses

Integration is a key component of mixed methods approaches, yet how integration of quantitative and qualitative methods occurs is a developing field. Integration of the quantitative and qualitative approaches in a research study can occur at various points in the course of a project. We have already seen, above, how integration is incorporated into the design of a study. In interpretation and reporting the results of a mixed methods study, several data analysis and display strategies can be employed to facilitate integration: narrative, data transformation, and joint displays [39]. Narrative discussion weaves a story based on the combined results of the quantitative and qualitative components organized theme-by-theme. In data transformation, one type of data is transformed into the other. An example would be counting themes derived from qualitative text analysis, transformation of the data into variables, and combining with the quantitative data for statistical analysis. Using joint displays, the qualitative themes are arrayed against quantitative data or characteristics. The juxtaposition of the data analysis in this way can lead to new insights [40].

In the context of our discussion of integration, the concept of "triangulation" should be mentioned. Triangulation refers to the idea that we may feel more confident in our findings if we arrive at the same conclusions with different methods (e.g., open-ended qualitative interviews yield the same findings as survey research). While triangulation may be a reason for using mixed methods, more often than not the use of qualitative methods alongside quantitative methods provides new insights into community or participant perspectives. Sometimes the most interesting findings can emerge from points where the methods appear to be leading to different conclusions. Such a realist or constructivist perspective allows for multiple viewpoints rather than an emphasis on uncovering a single 'true' perspective [41].

## 10.7 Examples of Studies Using Mixed Methods

In the remaining sections of the chapter, we would like to draw on our own research to provide examples of how mixed methods research has been used in studies of older persons: cultural factors in mental health and intervention development. These two broad areas illustrate the use of the principles discussed earlier in the chapter. To study cultural factors in depression among older persons, we make use of the simple designs discussed above. In considering intervention development, we discuss complex designs that use the simple designs as building blocks.

### *10.7.1 Cultural Factors in Mental Health*

Patients and providers are steeped in culture, which are patterns of thought and behavior that are learned and shared. Culture gives meaning and guides people in their daily lives. Cultural considerations need to take into account when understanding minority populations and specifically in the delivery of mental health services [42]. In mental health, professional providers use concepts that are distinctly Western to diagnose mental illness (an *etic* perspective). Standard concepts commonly included in depression assessments, such as anhedonia, and somatic symptoms like poor sleep and appetite, constitute the diagnosis of major depression. Older adults who have immigrated in late life may have lived most of their lives in other cultures and may have perceptions and ideas of depression that are different from Western concepts regarding how depression is identified and what should be done about it. In order to serve minority groups in a culturally competent and effective way, providers should be knowledgeable about different ideas of depression and conditions that restrict help seeking.

A mixed methods approach is useful in understanding underserved older adults who may be culturally different. We discuss a study that used an explanatory sequential mixed methods design with purposive sampling to study the meaning of depression among Korean older adults who had immigrated to the United States. A quantitative study, called Memory and Aging Study of Koreans in Maryland, (MASK-MD) was conducted first. MASK-MD was a descriptive epidemiological study among Korean elderly that assessed the prevalence of dementia and depression and mental health service use [43]. The study showed that only a small portion of Korean older adults with clinically significant depression accessed professional mental health treatment.

Quantitative analysis did not provide information on the potential causes of low use. In order to understand this finding, a qualitative study was conducted [44]. A subset of participants in the MASK-MD study who had clinically significant depression scores was selected for semi-structured interviews. The rationale for the sampling method included a desire to obtain the perspective of Korean elderly who were depressed, and so were knowledgeable about depression and barriers to services yet did not use services or receive any treatment.

The results of the study showed that Korean older adults described their depression in culturally-specific ways based on the perceived success or failure of their immigration experience and intergenerational conflicts with their children with whom the older adults shared a complex relationship. Social pressure and experiences of stigma served to suppress help seeking due to the real fear of social marginalization. Most important of all, participants desired services, but had few options for mental health services that were culturally relevant and provided in their language.

Quantitative studies can provide results that are generalizable; however, quantitative methods are based on assumptions and theories that may not reflect assumptions and theories of those populations they are trying to understand. Quantitative studies often yield outcomes regarding disease or service use that may be too abstract and general for direct application to specific local situations, contexts, and individuals, with limited use in developing interventions [1]. Mixed methods in this study explained the low use of mental health services found in the epidemiological study and provided an understanding of Korean older adults who may want services but cannot engage in services due to cultural barriers. Such knowledge is critical in designing effective and culturally relevant interventions.

### *10.7.2 Intervention Development*

Intervention development is a complex process that involves multiple stakeholders. Particularly in psychosocial interventions where behavior change is the focus, mixed methods is necessary to fully understand objective outcomes as well as the perspectives of those who deliver and receive the intervention. Psychosocial interventions typically include provision of services through the development of a relationship between the provider and the patient, and in this relationship, communication in the dyad is an important element that contributes to success or failure. We discuss an example in which mixed methods were used sequentially to develop an intervention, called Peer Enhanced Depression Care that used peer mentors to deliver depression care to underserved older adults.

Mixed methods were an integral part of intervention development from the beginning. The pilot study of Peer Enhanced Depression Care used a single cohort, pre-post design and used standard questionnaires to assess the outcome (depression) of the intervention. Patient-level factors that could serve as potential mediators such as loneliness, self-efficacy and coping skills were also measured. The peer-based intervention decreased depression and loneliness and increased coping skills for older adults in the sample [45, 46]. In order to understand the meaning of these outcomes from the perspective of the participants, qualitative methods were used in the form of semi-structured interviews at the post-study assessment to understand the strengths and benefits of peer-delivered depression care and the impact, if any, of participation in the study on depression for patients and peer mentors [45].

All participants (30 older adults and 6 peer mentors) were interviewed. Persons whose depression improved emphasized the importance of a trusting relationship, the credibility of the peer and usefulness of professional involvement in the program. Participants described benefits such as hope, changes in attitude, behavior and insight. Persons whose depression did not improve did not develop a trusting relationship with the peer. Peer mentors expressed the importance of training and professional supervision.

Given the importance of communication in relationship building between the peer mentors and patients, all interactions were audio recorded. With audio recordings, we conducted a quantitative study to assess peer communication to understand which peer communication behaviors increased rapport and decreased depression [46]. The goal was to use the results of the analysis of audio recordings to refine the peer mentor role and the intervention.

Audio-recordings were purposively sampled and analyzed using the Roter Interaction Analysis System, a quantitative method that is used to code frequencies of specific kinds of talk. The numeric results showed that self-disclosure was an important element in establishing rapport but was not associated with decreasing depression [46]. Self-disclosure is acknowledged as an important element in peer support, but it can have both positive and negative effects presumably based on what is said and how communication is expressed. Clearly training of peer mentors surrounding self-disclosure is an important element of intervention training and a nuanced understanding of the circumstances and effects of self-disclosure are needed.

In order to further understand use of self-disclosure by peer mentors beyond counting occurrences, the audio recordings from the study were analyzed qualitatively. Specifically, we reviewed the segments of the audio recordings where peer self-disclosure occurred and reviewed these segments with attention to the content of the self-disclosure as well as how self-disclosure was used in the interaction by the peer mentors. Qualitative study of the audio recordings showed that sometimes peer mentors self-disclosed information about themselves that was related to the conversation at a specific moment but not related to depression. Peer mentors also used self-disclosure to communicate similarity with the patient in a socially supportive manner; that is, that they had experienced the same thing as the patient to normalize an experience for a patient. Most significantly, peer mentors used self-disclosure to tell stories about themselves that communicated a lesson or coping strategy that could be useful for the patient specifically for reducing depression [47].

The mixing of quantitative and qualitative methods has been necessary for the development of this intervention. Quantitative methods were used to assess preliminary outcomes followed by qualitative methods to understand patient and peer perspectives and their experiences in the study. A quantitative health communication system helped elucidate which communication factors were specifically important for establishing rapport and decreasing depression, while qualitative review of peer communication revealed how a communication element was used.

## 10.8 Challenges and Opportunities in Mixed Methods Research

A challenge in using both quantitative and qualitative approaches in designing and testing interventions is to have sufficient time and resources allocated to ensure sufficient rigor for both approaches. Though a qualitative approach using participant interviews and observations may yield detailed information, sampling for qualitative approaches typically involves a small number of people (often less than 50). Such sample sizes typically preclude the use of statistical testing (e.g., of differences in participant characteristics according to themes that emerged in qualitative interviews [11]. We want to have representativeness of the main themes, perceptions, and insights (as we continue to talk to people, do new themes emerge or have we reached 'saturation' in that no new themes emerge?), as opposed to estimating a population parameter (which may depend on random selection of sample large enough to provide statistical power). Another challenge is the initiation of collaboration with investigators who are used to restricting their data collection strategies to quantitative methods to adequately allocate resources to collect and analyze qualitative data from observations, narratives, and visual data [48]. Fostering trans-disciplinary research teams for mixed methods behavioral health research poses a challenge [49]. In order to maximize the use of mixed methods in RCTs, principal investigators of clinical trials may benefit from including researchers and staff who are trained in qualitative and mixed method approaches, beginning at the study design stage [3]. Teams doing mixed methods research may benefit from time set aside for problem-solving and discussions about study design, underlying epistemological assumptions, and interpretations. Incorporation of qualitative methods in intervention studies, including randomized trials, can be associated with pitfalls that threaten the external validity of the trial [50]. Finally, in analysis of mixed data, findings from one strand may be contradictory or discordant with the other strand. While this may be a challenge or viewed as a weakness, such discrepancies can lead to new insights about the processes or measurements we might otherwise not question.

## 10.9 Conclusions

Mixed methods enable investigators conceptually and analytically to integrate qualitative research and qualitative data (e.g., semi-structured interviews, observations, focus groups) with traditional epidemiological and quantitative methods of research to facilitate translation of new interventions into practice. Mixed methods help understand, not just whether an intervention works, but how, why, and for whom. While quantitative approaches can characterize and measure patient outcomes, use of mixed methods can enhance quantitative analyses to identify unmeasured factors that might be associated with poor response to interventions [51], or factors that account for people who do not "fit the model," that is outliers [52], or provide clues to how interventions can incorporate the diverse needs and circumstances of people ("personalized interventions" [14]). Because mixed methods research designs place high value on the stories behind the numbers – both in exploratory designs where the experiences and insights of the community under study inform the quantitative investigation, and in explanatory designs where they illuminate the quantitative data – mixed methods are especially appropriate in studying the problems of older adults and caregivers.

## References

1. Johnson RB, Onwuegbuzie AJ, Turner LA (2007) Toward a definition of mixed method research. J Mixed Methods Res 1:112–134
2. Gallo JJ, Lee SY (2016) Mixed methods in behavioral intervention research. In: Gitlin LN, Czaja SJ (eds) Behavioral intervention research. Springer, New York, pp 195–211
3. Robins CS et al (2008) Dialogues on mixed-methods and mental health services research: anticipating challenges, building solutions. Psychiatr Serv 59(7):727–731. https://doi.org/10.1176/appi.ps.59.7.727
4. Pluye P, Hong QN (2014) Combining the power of stories and the power of numbers: mixed methods research and mixed studies reviews. Annu Rev Public Health 35(1):29–45. https://doi.org/10.1146/annurev-publhealth-032013–182440
5. Ivankova NV, Kawamura Y (2010) Emerging trends in the utilization of integrated designs in the social, behavioral, and health sciences. In: Tashakkori A, Teddlie C (eds) Sage handbook of mixed methods in social and behavioral research. Sage, Thousand Oaks, pp 581–611
6. Coyle CE et al (2018) Federal funding for mixed methods research in the health sciences in the United States: recent trends. J Mixed Methods Res 12(3):305–324. https://doi.org/10.1177/1558689816662578
7. Plano Clark VL (2010) The adoption and practice of mixed methods: U.S. trends in federally funded health-related research. Qual Inq 16:428–440
8. Nass P, Levine S, Yancy C (2014) Methods for involving patients in topic generation for patient-centered comparative effectiveness research: an international perspective. Patient-Centered Outcomes Research Institute, Washington, DC
9. Morse JM, Niehaus L (2009) Mixed method design: principles and procedures. Left Coast Press, Walnut Creek
10. Albright K, Gechter K, Kempe A (2013) Importance of mixed methods study in pragmatic trials and dissemination and implementation research. Acad Pediatr 13(5):400–407. https://doi.org/10.1016/j.acap.2013.06.010
11. Wittink MN, Barg FK, Gallo JJ (2006) Unwritten rules of talking to doctors about depression: integrating qualitative and quantitative methods. Ann Fam Med 4(4):302–309
12. Curry LA, Nembhard IM, Bradley EH (2009) Qualitative and mixed methods provide unique contributions to outcomes research. Circulation 119(10):1442–1452. https://doi.org/10.1161/CIRCULATIONAHA.107.742775
13. Farquhar MC, Ewing G, Booth S (2011) Using mixed methods to develop and evaluate complex interventions in palliative care research. Palliat Med 25(8):748–757. https://doi.org/10.1177/0269216311417919
14. Curry L, Shield R, Wetle T (eds) (2006) Improving aging and public Health Research: qualitative and mixed methods. American Public Health Association, Washington, DC
15. Nastasi BK et al (2007) Mixed methods in intervention research: theory to adaptation. J Mixed Methods Res 1(2):164–182
16. Betancourt T et al (2011) Using mixed-methods research to adapt and evaluate a family strengthening intervention in Rwanda. Afr J Trauma Stress 2(1):32–45
17. Bass JK et al (2013) Controlled trial of psychotherapy for Congolese survivors of sexual violence. N Engl J Med 368(23):2182–2191. https://doi.org/10.1056/NEJMoa1211853
18. Greenhalgh T et al (2010) Adoption and non-adoption of a shared electronic summary record in England: a mixed-method case study. BMJ 340:c3111. https://doi.org/10.1136/bmj.c3111
19. Bradley EH et al (2009) Research in action: using positive deviance to improve quality of health care. Implement Sci 4:25
20. Brannen J, Moss G (2012) Critical issues in designing mixed methods policy research. Am Behav Sci 56(6):789–801
21. Petros SG (2011) Use of a mixed methods approach to investigate the support needs of older caregivers to family members affected by HIV and AIDS in South Africa. J Mixed Methods Res 6:275–293

22. Apesoa-Varano EC, Hinton L (2013) The promise of mixed-methods for advancing latino health research. J Cross Cult Gerontol 28(3):267–282. https://doi.org/10.1007/s10823-013-9209-2
23. Stewart M et al (2008) Researching reducing health disparities: mixed-methods approaches. Soc Sci Med 66(6):1406–1417. https://doi.org/10.1016/j.socscimed.2007.11.021
24. Barg FK et al (2006) A mixed methods approach to understand loneliness and depression in older adults. J Gerontol B Psychol Sci Soc Sci 61(6):S329–S339
25. Curran GM et al (2012) Effectiveness-implementation hybrid designs: combining elements of clinical effectiveness and implementation research to enhance public health impact. Med Care 50(3):217–226. https://doi.org/10.1097/MLR.0b013e3182408812
26. Glasgow RE, Emmons KM (2007) How can we increase translation of research into practice? Types of evidence needed. Annu Rev Public Health 28:413–433
27. Glasgow RE et al (2012) National Institutes of Health approaches to dissemination and implementation science: current and future directions. Am J Public Health 102(7):1274–1281. https://doi.org/10.2105/AJPH.2012.300755
28. Palinkas LA (2014) Qualitative and mixed methods in mental health services and implementation research. J Clin Child Adolesc Psychol 43(6):851–861. https://doi.org/10.1080/15374416.2014.910791
29. Tashakkori A, Teddlie C (1998) Mixed methodology. Sage, Thousand Oaks
30. Creswell JW et al (2003) Advanced mixed methods research designs. In: Tashakkori A, Teddlie C (eds) Handbook of mixed methods in social and behavioral research. Sage, Thousand Oaks
31. Borkan JM (2004) Mixed methods studies: a foundation for primary care research. Ann Fam Med 2(1):4–6
32. Morgan DL (2007) Paradigms lost and pragmatism regained: methodological implications of combining qualitative and quantitative methods. J Mix Methods Res 1:48–76
33. Morse JM (2002) Qualitative tokenism. Qual Health Res 12(6):729–730. https://doi.org/10.1177/104973230201200601
34. Patton M (2003) Qualitative evaluation and research methods, 2nd edn. Sage, Newbury Park
35. Tashakkori A, Teddlie C (2003) Handbook of mixed methods in social & behavioral research. Sage, Thousand Oaks
36. Creswell JW, Plano Clark VL (2011) Designing and conducting mixed methods research, 2nd edn. Sage, Washington, DC
37. Weller N, Barnes J (2014) Pathway analysis and the search for causal mechanisms. Sociol Methods Res 3(45):424–457
38. O'Cathain A et al (2014) Getting added value from using qualitative research with randomized controlled trials: a qualitative interview study. Trials 15:215. https://doi.org/10.1186/1745-6215-15-215
39. Fetters MD, Curry LA, Creswell JW (2013) Achieving integration in mixed methods designs-principles and practices. Health Serv Res 48(6 Pt 2):2134–2156. https://doi.org/10.1111/1475-6773.12117
40. Guetterman TC, Fetters MD, Creswell JW (2015) Integrating quantitative and qualitative results in health science mixed methods research through joint displays. Ann Fam Med 13(6):554–561. https://doi.org/10.1370/afm.1865
41. Maxwell JA (2011) A realist approach for qualitative research. Sage, Thousand Oaks
42. Marsella A, Yamada A (2000) Handbook of multicultural mental health assessment and treatment of diverse populations. Assessment and treatment of diverse populations. Academic Press, San Diego
43. Lee HB et al (2014) Mental health service utilization among Korean elders in Korean churches: preliminary findings from the memory and aging study of Koreans in Maryland (MASK-MD). Aging Ment Health 18(1):102–109. https://doi.org/10.1080/13607863.2013.814099
44. Lee-Tauler SY et al (2016) What does depression mean for Korean American elderly?: a qualitative follow-up study. Psychiatr Investig 13(5):558–565
45. Joo JH et al (2016) An innovative model of depression care delivery: peer mentors in collaboration with a mental health professional to relieve depression in older adults. Am J Geriatr Psychiatry 24(5):407–416. https://doi.org/10.1016/j.jagp.2016.02.002

46. Joo JH et al (2017) The impact of peer mentor communication with older adults on depressive symptoms and working alliance: a pilot study. Patient Educ Couns S0738-3991(17):30588–30588. https://doi.org/10.1016/j.pec.2017.10.012
47. Truong C et al (2018) The role of self-disclosure by peer mentors in the delivery of depression care to underserved older adults. IN PREPARATION
48. George DR (2011) Intergenerational volunteering and quality of life: mixed methods evaluation of a randomized control trial involving persons with mild to moderate dementia. Qual Life Res 20(7):987–995. https://doi.org/10.1007/s11136-010-9837-8
49. Kessel F, Rosenfeld PL, Anderson NB (eds) (2008) Interdisciplinary research: case studies from health and social science, 2nd edn. New York, Oxford University Press
50. Cooper C et al (2014) Conducting qualitative research within clinical trials units: avoiding potential pitfalls. Contemp Clin Trials 38(2):338–343. https://doi.org/10.1016/j.cct.2014.06.002
51. Kravitz RL, Duan N, Braslow J (2004) Evidence-based medicine, heterogeneity of treatment effects, and the trouble with averages. Milbank Q 82(4):661–687. https://doi.org/10.1111/j.0887-378X.2004.00327.x
52. Kawachi I, Berkman LF (2003) Neighborhoods and health. Oxford University Press, Oxford

# Chapter 11
# Systematic Reviews and Meta-Analysis in Aging Research

**Miguel Ángel Villasís-Keever, Mario Enrique Rendón-Macías, and Raúl Hernán Medina-Campos**

**Abstract** Systematic reviews and meta-analyses are proven tools for decision-making in health care, both for patients and public policy. For example, nowadays they constitute a substantial part of evidence-based clinical practice guidelines. However, the number of systematic reviews developed so far, and their use to improve the health of older adults has been somehow slow. This chapter describes in detail each of the steps necessary to conceptualize and conduct systematic reviews and meta-analysis. It begins with a description of the different uses these types of tools have, followed by the differences they have with narrative reviews. Regarding the methodology to assemble them, it starts in the form of how the research question is formulated, which is the essence for the construction of each of systematic reviews. Then we continue with the selection of studies, first by searching in different electronic databases (e.g., Medline, Embase). Once studies are located, each of them should be reviewed thoroughly to determine if they comply strictly with the selection criteria. Finally, with the selected studies the next step is data extraction from each one, which eventually constitutes the results section of the systematic review. In addition, it is necessary to assess the methodological quality of each study to determine if they are free of bias. The last part of the chapter focuses on the different alternatives of meta-analyses, including network meta-analysis. The results are reported qualitatively when they are systematic reviews, while meta-analyses are reported quantitatively, as long as two or more studies can be combined.

**Keywords** Systematic review · Meta-analysis · Network meta-analysis · Secondary research

---

M. Á. Villasís-Keever (✉) · M. E. Rendón-Macías
Research Unit in Clinical Epidemiology, Pediatrics Hospital, National Medical Center Century XXI, Mexican Institute of Social Security, Mexico City, Mexico
e-mail: miguel.villasis@imss.gob.mx; mario.rendon@imss.gob.mx

R. H. Medina-Campos
Department of Geriatric Epidemiology, National Institute of Geriatrics, Mexico City, Mexico
e-mail: rmedina@inger.gob.mx



C. García-Peña et al. (eds.), *Aging Research - Methodological Issues*,
https://doi.org/10.1007/978-3-319-95387-8_11

## 11.1 Introduction

Finding efficient and effective methods for the care of older adults is important. However, it is often the case that the skills to find, evaluate critically and synthesize the information of published studies to integrate the results of the research to normal clinical practice are not available. In this context, several years ago systematic reviews and meta-analyses have been growing in numbers as a tool for making decisions in different areas of medicine, from the moment the doctor meets the patient through conception of a research project or the planning of health-care policies. However, its use has been slow to seep into every possible area, possibly due to lack of knowledge of its scopes, advantages and reliability [1]. In part, this type of research has emerged because the volume of information is growing every day, and this makes it necessary to make documents available that summarize in orderly fashion and under scientific criteria the state of knowledge of a given topic, in order to facilitate decision-making for health-care personnel, researchers, patients and those who implement health-related public policies. From the start of the development of the process of an orderly summarizing or synthesizing and criticizing scientific evidence until arriving at what today is known as systematic review and meta-analysis, different terms have been used: review articles, synthesis of research, overview, etc. As well, various definitions have been developed, but one of the most widely accepted is the one proposed by Ian Chalmers and Douglas Altman – two of the main pioneers in this area – which establishes that: "a review is what has been prepared through a systematic process to minimize biases and random errors, which is documented in the material and methods section" [2]. With this definition, it is clear that a systematic review is a research project that includes studies from a single theme or topic [3]. Meta-analysis is the term used when a statistical analysis of the results of at least two individual studies obtained from a systematic review is carried out. Thus, it is necessary to recognize that all meta-analyses are also systematic reviews. Systematic reviews are also known as secondary investigations, since they deal with synthesizing the information from primary or original studies that have been done through the years on a particular theme. In other words, a systematic review concentrates all of the studies carried out on a specific theme using the scientific method [4]. Although earlier systematic reviews and meta-analyses focused on synthetizing studies on therapeutic interventions (like drugs or surgical interventions) described in controlled clinical trials to determine which were the most effective, over time the spectrum has broadened to include observational studies whose objectives are to learn the prognosis of illnesses, evaluate the best diagnostic tool, or establish causes or risk factors of illnesses [5].

## 11.2 Types of Reviews

In spite of the development and evolution this field of research has had in recent decades, not all "review" articles or publications are systematic reviews. Because of this, it is necessary to distinguish between systematic reviews and narrative reviews. Narrative reviews are documents that describe a specific theme or topic, which may be considered similar to what is usually presented in a chapter of a book, where historic aspects, epidemiology, signs and symptoms, or the clinical course of an illness are presented, as well as the process of making the diagnosis and even the therapeutic options [6]. In the case of systematic reviews, although they also deal with a particular illness, the approach is very specific; in other words, the synthesis of the information is centered only on the treatment, diagnosis, risk factors or prognosis [7]. There are also differences in the process of including the various articles or references that support the authors' arguments; in a narrative review there is no selection process, even though it may be based on recent publications. For this reason, it is thought that the information in these documents is directed specifically toward the concepts or proposals of the authors themselves, since the support documents only agree with what is described in the text and do not include everything published on the theme. For example, if the author of a narrative review has been distinguished for having carried out research on a particular drug, then possibly the information included in the article would be limited specifically to the findings that he or she has described, even if there may have been other studies with different results. In contrast, in systematic reviews where a search of all the available studies in world literature is carried out, it will include articles both in favor of and against the effectiveness of the drug, as well as all the adverse events. As well, if possible, a systematic review could include what is called gray literature, that is, the results of research not published in scientific periodic circulation journals, such as theses or works presented at medical conferences [8].

## 11.3 Uses of Systematic Reviews

Because systematic reviews synthesize research studies published up to the present, they are one of the basic tools in Evidence-Based Medicine (EBM) [8]. As we know, two of the principal steps of EBM refer as much to the search as to critical reading of the studies, which will support medical decisions for patients on aspects mainly related to the diagnosis, treatment or prognosis. By virtue of the fact that a research group has previously completed the task of identification and evaluation of published studies, the use of systematic reviews and meta-analysis will be beneficial for the practice of EBM. For example, in the case where the review might have demonstrated the effectiveness of a drug for the treatment of hypertension in patients older than 65 years of age, after reading a systematic review of controlled clinical trials published up to the present, the doctor, instead of reviewing and analyzing each clinical trial where a particular drug was analyzed, could direct his/her efforts toward determining the viability of giving this intervention to his/her patients (for availability, cost or the characteristics of the patients), or evaluating the result once the intervention is implemented [6]. Moreover, at present systematic reviews are essential for preparing evidence-based clinical practice guidelines. An example is where more than one option is available for aspects of diagnosis or treatment, in which case the recommendations of the guidelines will be based mainly on the results of the studies with the best quality. In this context, as the treatment of hypertension involves a therapeutic intervention, if a meta-analysis of randomized controlled clinical trials is identified where it is shown that this intervention is effective and with few side effects for patients older than 65 years, then that information would be included in the guideline, and would form part of the recommendations [9]. Another use that has arisen in recent years, both of systematic reviews and meta-analyses, is in health-care policy, where they are becoming increasingly important as a reference point. For example, in making decisions on the incorporation of new drugs for a specific illness, whether for a hospital or health system, all of the alternative therapies available are considered, including the results of their clinical effectiveness and costs. In this latter category it must be emphasized that publication of economic studies is limited; still, for experts in health economics systematic reviews are extremely useful for identifying the impact of each of the interventions in health outcomes (therapeutic efficiency, morbidity, mortality, quality of life, to name a few), and this could serve to bring about cost-effectiveness, cost-utility and cost- benefit evaluations [5].

## 11.4 The Process of Preparing Systematic Reviews

Like all research, systematic reviews must start with the drawing up of a protocol. This protocol will describe the process for executing each of the steps with the smallest number of errors or biases. Fig. 11.1 outlines the process.

### *11.4.1 Preparation of the Research Question*

The first step in doing a systematic review is to focus as much as possible on studying a single fundamental, answerable question by constructing one that determines the scope of the study [10]. The more concrete the question is the easier will be the review. To construct the question the following must be taken into account: (1) the population of the study, where it includes age (children, adults, older adults) and the illness to be studied; in other words, the illness itself or one or more of its complications; (2) the intervention to be evaluated, whether it be treatment, diagnosis, risk factors or prognosis; (3) the possible options for intervention; for example, two or more drugs, procedures, diagnoses, etc.; (4) the different results of the effect of the

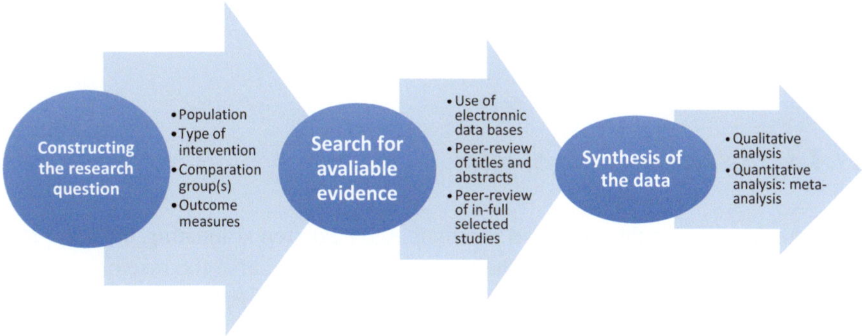


**Fig. 11.1** The process for drawing up a systematic review

intervention, for example, if it is therapeutic, then the results to evaluate are to avoid complications or death; while if it is diagnosis, the capacity of one or more tests to confirm or rule-out the illness being studied.

### *11.4.2 The Search for Articles*

Once the research question has been defined, the next step is to start the search for studies that have been done on the theme. To begin this task, the researcher has to take into account each of the elements included in the question. To do this, electronic databases are used that concentrate the bibliography or publications on health care themes. The largest databases in the world are Medline and EMBASE, which use terms or text words (also known as keywords) to classify each publication. Other important sources could be the Cochrane Collaboration and CINAHL. There are now proven methods for obtaining the most efficient searches, in other words, which have the highest sensitivity for locating pertinent studies to resolve research questions [10]. However, one problem in this search is that in these databases, only the titles and the summaries of the publication are available, so it is necessary to read and scrutinize to choose the articles that would potentially serve for the systematic review and then get their full version. In a high-quality systematic review, the selection of titles and summaries is done by a peer-review process, which means that two or more researchers work together to discriminate and decide which studies meet the inclusion criteria. The goal in this phase is to choose only the articles where both researchers agree to include them; if there is disagreement, they could consult a third researcher, who could decide on whether or not to include it, or they could make a consensual decision [11]. When doing the search for summaries on these databases, articles in any language could be identified, so it is important to decide whether language restrictions will be included in the criteria. In this context, the reader of a systematic review should be aware of the possibility of biases about this. However, when the inclusion of studies is restricted to English, it is seen as a small error, since about 90% of world scientific production is published in that language.

### 11.4.3 Selection of the Studies

After having chosen the titles and abstracts, the next step is locating the complete versions of the studies selected. Once the studies are available the researchers proceed to review them for each of the components (especially in the methodology and results sections) to determine whether they effectively comply with each point identified in the research question. As one would expect, it is only at this time that it can be determined with certainty whether the studies contain enough and adequate information to be incorporated into the systematic review and subsequent analysis. This is another phase of scrutinizing, since a large proportion of the articles recovered will not contain the minimum elements necessary to be analyzed. For example, in systematic reviews of therapeutic interventions, only randomized controlled clinical trial must be included. In addition, it is necessary to have clear data about the results of the interventions in evaluation. On this point, it should be mentioned that there are publications that refer to research protocols, reviews, or initial reports, but without data, so these articles must be excluded. Added to the foregoing and with the goal of identifying a larger number of publications that would not have been located with the search strategy, in many systematic reviews the reviewers opt to review the list of bibliographical references of the studies included [10]. Lastly, the quality of this phase is again guaranteed when a peer-review process is carried-out, in the same way the summaries were selected.

### 11.4.4 Data Extraction

In this phase of the systematic reviews, the researchers try to obtain useful information contained in each of the studies that will conform to the final analysis and that complies with the study's objectives. To do this, the researchers must set up data collection sheets on each variable, taking into account all the variables: outcomes and intervention, as well, as demographic data. Based on what is described in each of the studies included, these sheets will be filled with information on the general characteristics of each study, where (e.g., country or a hospital) it was done, the characteristics of the patients included (age, sex, selection criteria of the patients, number of participants, severity of disease, etc.), the intervention (for example, studies on therapy should specify doses, methods of administration, duration time of the intervention, co-interventions) and, of course, the results (outcome measures) of the way in which the original article shows them. If necessary authors of systematic reviews will have to include the definitions used in each study, since these could vary [8]. In this phase, it is recommended that the information be extracted and recorded by a peer-review process, as mentioned for the previous steps [5].

### *11.4.5 Quality Assessment of Included Studies*

Analysis of methodological quality of each included study is an important aspect in all systematic reviews. It is necessary to make an evaluation of how each study was carried out, according to what was described by the authors. For example, in a clinical trial both randomization, blindness of measurements and complete follow-up of all participants are essential to determine that a study is valid. Studies with higher quality, results will be more reliable. When there are problems in the execution of clinical trials, that is, with lower quality or bias, the benefits of an intervention are generally overestimated [10].

### *11.4.6 Analyses of the Results of a Systematic Review*

This phase of the systematic reviews depends on the findings obtained in each of the previous phases. Thus, to determine the way the results will be described, it is necessary to know whether or not the studies included were similar. This is done by means of an individual analysis of each one, contrasting it with the others. It is important to evaluate the selection criteria of patients, the intervention, and the method of measuring the variables, whether they were similar among the studies [11]. If differences in each of these principal aspects are observed, it is considered that there is heterogeneity among the studies, and thus the results will be made only on the qualitative aspect. This could mean that the final report will be based on the description of the characteristics of each of the studies included (qualitative analysis). This is known as systematic review [2].

When the studies are considered to be similar, that is, there is homogeneity; a statistical analysis could be carried out combining the data on one or more variables of results when these come from at least two different studies. This procedure is known as meta-analysis. To do this it is necessary to consider the scales of measurement of the variables used by the authors of the studies included, since the relevant statistical method will be selected in accordance with their nature. For example, if the review deals with evaluating the therapeutic effect of a drug for hypertension, it is possible that the outcome measures have been established as disappearance of the hypertension (a variable with qualitative measurement scale) or modification of the blood pressure values (a variable with a quantitative measurement scale). Each variable will be analyzed with different estimates, which will be obtained by separate statistical procedures (see below). At present there are several statistical packages available, such as RevMan o Metaanalyst, for working with these specific data in carrying out a meta-analysis.

### 11.4.7 Final Report and Recommendations

The last step in systematic reviews is the preparation of the final report, which, in addition to describing the entire process carried out, includes the results obtained from the qualitative or descriptive perspective and, if relevant, the quantitative results when the meta-analysis was done. It is important when preparing the final report to be objective and neutral on the findings, and to provide information on all the elements so that future readers can make their own decisions [4]. It is not rare to find systematic reviews where the authors indicate the existence of not enough evidence to achieve a definitive conclusion because of the poor quality of the published studies, or due to a small number of patients studied, or because the available studies have not shown obvious clinical benefits.

## 11.5 Meta–Analysis

To deepen the theme of meta-analysis, it should be mentioned that one of the most important reasons for doing this type of research, was to think about the possibility of increasing the number of subjects studied on a specific topic obtained by adding up the results of different studies. In this regard, a number of studies worldwide, it is common the authors recognize that due to the small size of their study, they did not dare to offer a solid conclusion about their findings, for example, whether a new intervention is better than the usual treatment. When a meta-analysis is done it is accepted that it will be equivalent to having done a single study but with a much larger sample size, in other words with more statistical power, since it combines the results from two or more studies to issue a more reliable conclusion. Considering this scenario, by increasing the number of participants it is more probable that it will demonstrate the true effectiveness of an intervention. If this assumption is true, it is feasible to make decisions based on studies that are already done without undertaking the task of doing a large-scale study. However, if there are still doubts despite these benefits of meta-analysis, nowadays it is preferred to do a study with a large sample size to determine the effectiveness of an intervention, since this ensures that all the participants in the study were exposed to similar maneuvers during the same period of time [9]. Another point to consider in carrying out a meta-analysis, as was mentioned previously, is the method of reporting the results, that is, the scale of measurement of the variables of the outcomes. The five most used measures are, for quantitative variables: the weighted mean difference and the standardized mean difference (the latter is also known as size of the effect) (Fig. 11.2); and for categorical variables: relative risk (RR), odds ratio (OR) and risk difference (Fig. 11.3) [12].

In general, there are two models for carrying out a meta-analysis, the fixed effects model and the random effects model. In the former, it is accepted that the studies included estimate the same "true" value of the effect, and that the differences observed among them are due to chance. In the random effects model it is assumed

| | Presente | | | Ausente | | | | Mean Difference |
|---|---|---|---|---|---|---|---|---|
| Study or Subgroup | Mean | SD | Total | Mean | SD | Total | Weight | IV, Fixed, 95% CI |
| Conter, 2010 | 41.8 | 8.1 | 38 | 86.9 | 0.9 | 1711 | 5.5% | -45.10 [-47.68, -42.52] |
| Escherich, 2010 | 53 | 15.5 | 11 | 85.4 | 1.5 | 566 | 0.4% | -32.40 [-41.56, -23.24] |
| Gaynon, 2010 | 45.5 | 7.2 | 63 | 86.9 | 0.8 | 2008 | 11.6% | -41.40 [-43.18, -39.62] |
| Kamps, 200 | 42.9 | 18.7 | 7 | 83.7 | 1.3 | 460 | 0.2% | -40.80 [-54.65, -26.95] |
| Liang, 2010 | 71.1 | 12.8 | 16 | 82.6 | 1.7 | 716 | 0.9% | -11.50 [-17.77, -5.23] |
| Michell, 2010 | 43.9 | 7.6 | 43 | 78.1 | 1 | 1791 | 7.1% | -34.20 [-36.47, -31.93] |
| Moricke, 2010 | 39.5 | 7.5 | 43 | 88.6 | 0.7 | 1917 | 7.3% | -49.10 [-51.34, -46.86] |
| Pui, 2010 | 50 | 17.7 | 6 | 84.3 | 2.9 | 159 | 0.2% | -34.30 [-48.47, -20.13] |
| Salzer, 2010 | 34.8 | 4.3 | 128 | 87.3 | 0.5 | 5228 | 65.8% | -52.50 [-53.25, -51.75] |
| Shmiegelow, 2010 | 78 | 14 | 9 | 89 | 1 | 228 | 0.4% | -11.00 [-20.15, -1.85] |
| Silverman, 2010 | 0 | 0 | 1 | 89.8 | 1.4 | 490 | | Not estimable |
| Stark, 2010 | 55.6 | 16.6 | 16 | 83.6 | 2.4 | 255 | 0.6% | -28.00 [-36.14, -19.86] |
| Tsuchida | 41.7 | 101 | 24 | 86.8 | 1.4 | 573 | 0.0% | -45.10 [-85.51, -4.69] |
| **Total (95% CI)** | | | **405** | | | **16102** | **100.0%** | **-48.43 [-49.03, -47.82]** |

Heterogeneity: Chi² = 570.55, df = 11 (P < 0.00001); I² = 98%
Test for overall effect: Z = 157.00 (P < 0.00001)

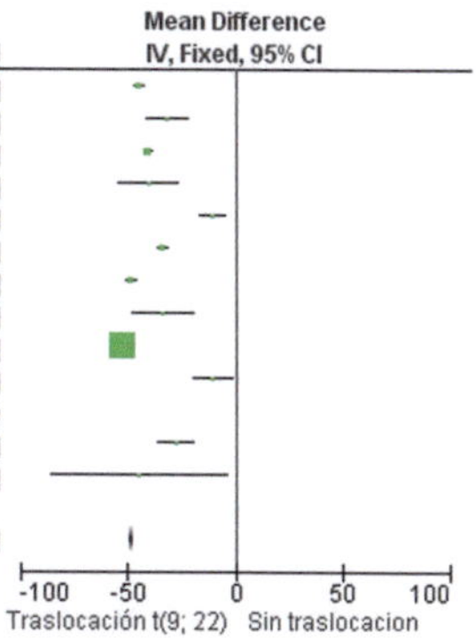


**Fig. 11.2** Representation by Forest Plot of meta-analysis with quantitative variables using weighted mean difference

| | Intervention | | Placebo | | | Odds Ratio |
|---|---|---|---|---|---|---|
| Study or Subgroup | Events | Total | Events | Total | Weight | M-H, Fixed, 95% CI |
| Barton, 2002 | 15 | 38 | 33 | 39 | 8.0% | 0.12 [0.04, 0.35] |
| Bikhazi, 2004 | 5 | 41 | 26 | 44 | 8.9% | 0.10 [0.03, 0.29] |
| Daniels, 2001 | 9 | 51 | 46 | 51 | 15.3% | 0.02 [0.01, 0.08] |
| Desjardins, 2001 | 4 | 54 | 37 | 54 | 13.8% | 0.04 [0.01, 0.12] |
| Hubbard, 2003 | 11 | 67 | 30 | 63 | 10.4% | 0.22 [0.10, 0.49] |
| Malan, 2003 | 1 | 55 | 7 | 61 | 2.6% | 0.14 [0.02, 1.20] |
| Mehlish, 2003 | 12 | 51 | 47 | 50 | 14.7% | 0.02 [0.01, 0.07] |
| Snabes, 2007 | 18 | 204 | 64 | 194 | 24.2% | 0.20 [0.11, 0.35] |
| Tang, 2002 | 2 | 17 | 6 | 18 | 2.1% | 0.27 [0.05, 1.57] |
| **Total (95% CI)** | | **578** | | **574** | **100.0%** | **0.11 [0.08, 0.15]** |
| Total events | 77 | | 296 | | | |

Heterogeneity: Chi² = 24.29, df = 8 (P = 0.002); I² = 67%
Test for overall effect: Z = 13.33 (P < 0.00001)

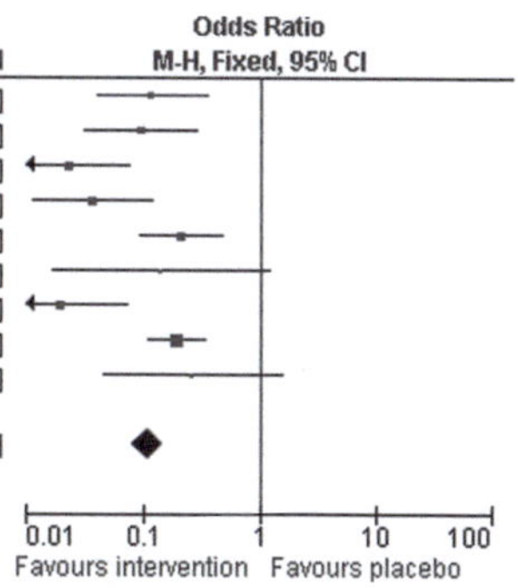


**Fig. 11.3** Representation by Forest Plot of a meta-analysis with qualitative variables using the odds ratio (OR) calculation

that the trials or studies included are only random samples from a "universe of studies," and that their results are positioned randomly around a single central value [11]. For presenting the results of the meta-analyses it is always useful to include graphs, with the forest plot being the most commonly used one, where in addition to describing the individual results from each study included, it shows the global estimate, which is considered the truest one. As shown on the right side of Figs. 11.2 and 11.3, a vertical line divides the graph, where the left side usually shows the benefit of the experimental intervention and the right side the control intervention. As well, each horizontal line represents each of the studies; this line has two components: (1) the estimation point (square) and (2) the extremes of the interval of reliability (horizontal line). In the lowest part of the graph is shown the global synthesis of all the studies (indicated with a diamond). As well, in all meta-analyses the differences between the studies from a statistical point of view are determined, which will indicate whether there is heterogeneity. The most frequent analyses for documenting this are the Breslow-Day, Q of Cochran, and Chi-squared tests. When

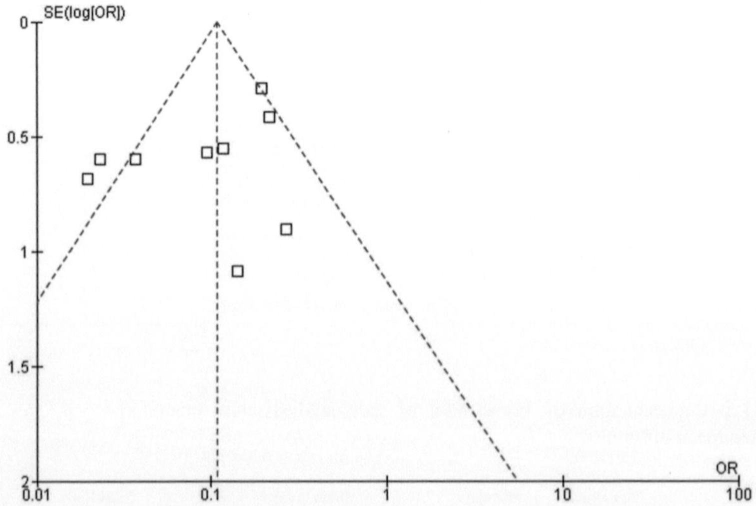


**Fig. 11.4** Example of a *funnel plot*

a value of $p \geq 0.10$ is obtained, it shows that there is heterogeneity, from which it can be inferred that there are important differences between the studies. The two main sources of heterogeneity are the number the participants in each included study (studies with many patients and others with few), and the discrepancy of results between studies. The latter is more often; because the effect of the intervention is not consistent, some studies indicate benefits and others do not. One strategy for resolving heterogeneity is to select a meta-analysis model of random effects or to carry out a sensitivity analysis. Sensitivity analysis could be done by analyzing only data from the studies with the largest number of participants, or only with those with the best quality, from the methodological point of view [1, 12].

Lastly, one aspect that must be mentioned is the possibility of a "publication bias." As one of the primordial objectives of systematic reviews and meta-analyses is to look for and concentrate all the studies on a specific theme in one study, it is necessary to take into account the tendency to publish only the studies that show benefits toward a determined intervention or manoeuver. Therefore, it is more difficult to find studies where conclusions go against the hoped-for effects "where no statistically significant differences are found." This has been named publication bias, since all meta-analysis should evaluate its possible existence [9]. The usual way of doing that is through graphic representation of this phenomenon. Graphs called funnel plots are the most commonly used to show the relationship between the size of the sample and the effect of the intervention from each study included. In this graph, a point represents each study. When there are no publication biases, the graph should appear as an inverted funnel which must contain all the points, but if parts of the funnel are lacking (absence of points or outside of the funnel), then publication bias might exist. If the later occur, the results should be interpreted with caution. A funnel plot is shown in Fig. 11.4, which shows that there may be publication bias.

## 11.6 Network Meta-Analysis

A frequent problem when performing meta-analysis of clinical trials is the limited number of studies comparing all available treatments. For example, there may be studies with a treatment A that is compared to a treatment B, but there are no studies that contrast treatment A with a C. However, treatment C has been contrasted with treatment B. In many occasions the reason for this comparison of treatments A and C with B, has been because the latter was the oldest to be approved in clinical use, although it may be the least effective. Both treatments A and C, through meta-analysis, have been shown to be more effective than treatment B when making direct comparisons. However, the question that remains is, which is more efficient or if its effectiveness is the same? This comparison would allow generating a more sustained recommendation.

The meta-analysis network is a method now available to estimate the comparative effectiveness between treatments that have not been directly contrasted. The estimate is calculated through the available studies of direct comparisons between related treatments. The result is an "indirect" estimate but close to reality, of the comparative effect between treatments not previously evaluated, one versus the other [13].

From mathematical point of view, this estimate is calculated by the principle of transitivity, where a measurement between two parameters not measured is obtained by the information available from these two parameters with a common comparator. The estimate of the effect C against A (indirect) = (direct effect of A against B) - (direct effect of C against B). However, to make this estimate valid, it is necessary to fulfill some conditions between the parameters:

1) Treatment B should be similar (doses, periods, etc.) when compared with treatments A and C. It is common that this is not fully complied, so some flexibility can be accepted. 2) The trials of each of the treatments compared with B must have been done with similar patients, if these differ substantially the final estimate will be inadequate. 3) It is required that patients have been randomly selected to any treatment. If this is not the case, confounding factors could cause errors in the results. 4) Undoubtedly, the efficacy criteria must have been evaluated in the same way in each of the included trials.

In addition, a network meta-analysis may compare an indirect estimator with a direct one. They are called mixed estimators, which serve to confirm the models (consistency), as well as to estimate the impact of the treatments under different conditions.

The reports of the results are the same as the classical meta-analysis. However, authors should clearly report (preferably in a graphic) comparisons made, in order to determine which were direct, indirect and mixed, also should be noted the number of included studies for each comparison [14].

Finally, we should remember that the best way to contrast two or more treatments is through direct comparisons, in well-conducted clinical trials.

## 11.7 The Quality of Systematic Reviews

It should be mentioned that, like all research processes, systematic reviews could have problems during their execution, for which the readers of this type of document must determine their reliability. To facilitate their evaluation, some instruments or scales have been developed over time to evaluate the quality of systematic reviews. These scales are based on the analysis of each of the steps (phases) of the preparation of a systematic review. As well, efforts have been made in which the publications of the reviews follow a standard format, called QUORUM, and more recently PRISMA [2, 5, 15, 16] to help readers evaluate them.

The items that evaluate these and other instruments are based on the good execution of the systematic reviews, most of which were mentioned above. It should be said that independently of whether or not the reader knows the essential elements for establishing whether a systematic review is of high quality, there are organizations that prepare them on a state-of-the-art basis, and these are usually reliable. The largest organization, and the one that sets the standards of quality for systematic reviews at present, is the Cochrane Collaboration. The advantages of this organization is that it is non-profit, and so all the reviews are available for consultation free of charge, and in many cases in different languages, including Spanish (www.bibliotecacochrane.com). Other organizations with reliable systematic reviews are those produced by the Agency for Healthcare Research and Quality (www.ahrq.gov) in the United States of America, or by Health Technology Assessment Programme in the United Kingdom (www.hta.ac.uk).

## References

1. Villasís-Keever MA (2000) Medicina Basada en la Evidencia. In: Novales J (ed) Medicina Interna Pediátrica. McGraw-Hill, México, pp 389–402
2. Moher D, Liberati A, Tetzlaff J, Altman DG, Group P (2009) Preferred reporting items for systematic reviews and meta-analyses: the PRISMA statement. PLoS Med 6(7):e1000097. https://doi.org/10.1371/journal.pmed.1000097
3. Egger M, Smith GD, Altman DG (2001) Systematic reviews in health care: meta-analysis in context, 2nd edn. BMJ, London
4. Song F, Parekh S, Hooper L, Loke YK, Ryder J, Sutton AJ et al (2010) Dissemination and publication of research findings: an updated review of related biases. Health Technol Assess 14(8:iii). ix–xi):1–193. https://doi.org/10.3310/hta14080

5. Moher D, Cook DJ, Eastwood S, Olkin I, Rennie D, Stroup DF (1999) Improving the quality of reports of meta-analyses of randomized controlled trials: the QUOROM statement. Quality of reporting of meta-analyses. Lancet 354(9193):1896–1900
6. Klassen TP, Jadad AR, Moher D (1998) Guides for reading and interpreting systematic reviews: I. Getting started. Arch Pediatr Adolesc Med 152(7):700–704
7. Ferreira Gonzalez I, Urrutia G, Alonso-Coello P (2011) Systematic reviews and meta-analysis: scientific rationale and interpretation. Rev Esp Cardiol 64(8):688–696. https://doi.org/10.1016/j.rec.2011.03.027
8. McMichael C, Waters E, Volmink J (2005) Evidence-based public health: what does it offer developing countries? J Public Health (Oxf) 27(2):215–221. https://doi.org/10.1093/pubmed/fdi024
9. Stroup DF, Berlin JA, Morton SC, Olkin I, Williamson GD, Rennie D et al (2000) Meta-analysis of observational studies in epidemiology: a proposal for reporting. Meta-analysis Of Observational Studies in Epidemiology (MOOSE) group. JAMA 283(15):2008–2012
10. Jadad AR, Moher D, Klassen TP (1998) Guides for reading and interpreting systematic reviews: II. How did the authors find the studies and assess their quality? Arch Pediatr Adolesc Med 152(8):812–817
11. Moher D, Jadad AR, Klassen TP (1998) Guides for reading and interpreting systematic reviews: III. How did the authors synthesize the data and make their conclusions? Arch Pediatr Adolesc Med 152(9):915–920
12. Thompson SG, Pocock SJ (1991) Can meta-analyses be trusted? Lancet 338(8775):1127–1130
13. Lumley T (2002) Network meta-analysis for indirect treatment comparisons. Stat Med 21(16):2313–2324
14. Bafeta A, Trinquart L, Seror R, Ravaud P (2014) Reporting of results from network meta-analyses: methodological systematic review. BMJ 348:g1741. https://doi.org/10.1136/bmj.g1741
15. Welch V, Petticrew M, Petkovic J, Moher D, Waters E, White H et al (2016) Extending the PRISMA statement to equity-focused systematic reviews (PRISMA-E 2012): explanation and elaboration. J Clin Epidemiol 70:68–89. https://doi.org/10.1016/j.jclinepi.2015.09.001
16. Hutton B, Salanti G, Caldwell DM, Chaimani A, Schmid CH, Cameron C et al (2015) The PRISMA extension statement for reporting of systematic reviews incorporating network meta-analyses of health care interventions: checklist and explanations. Ann Intern Med 162(11):777–784. https://doi.org/10.7326/M14-2385

# Chapter 12
# Health Systems Research in Aging

**Hortensia Reyes-Morales, Svetlana V. Doubova, and Ricardo Pérez-Cuevas**

**Abstract** As a relatively new area in health sciences, health systems research (HSR) is an interdisciplinary field that seeks to understand and improve how societies organize themselves for achieving health goals, and how health systems respond and adapt to health policies. HSR focusses on generating and transferring new knowledge that enables health programs, interventions and services, to respond effectively to health population needs. For aging population, health needs should consider a broader spectrum that comprises social determinants of health, and coordination among public sectors such a social networks, pensions, housing, and social services, among others, for improving their wellbeing. From this perspective, contribution of HSR to achieve suitable health systems and health services for the aging is particularly relevant. This chapter provides a general description of study designs and methods in HSR; it is intended to be descriptive and able to give an appropriate springboard for those who plan to conduct a type of research that has multiple and different layers, for contributing with scientific evidence in improving health in old people.

**Keywords** Health systems research · Health policy · Health systems

## 12.1 Concepts of Health Systems Research

Traditionally focused on biomedical or clinical research, scientific advances in health have been remarkable during the last century, and its contribution of cumulative knowledge for improving global health is unquestionable. Identification of

H. Reyes-Morales (✉) · R. Pérez-Cuevas
National Institute of Public Health, Cuernavaca, Morelos, Mexico
e-mail: hortensia.reyes@insp.mx; ricardo.perezcuevas@insp.mx

S. V. Doubova
Epidemiology and Health Services Research Unit "Centro Médico Nacional Siglo XXI", Mexican Institute of Social Security, Mexico City, Mexico
e-mail: svetlana.dubova@gmail.com



C. García-Peña et al. (eds.), *Aging Research - Methodological Issues*,
https://doi.org/10.1007/978-3-319-95387-8_12

diseases etiology, development of vaccines, innovative diagnostic methods or treatments, are only some examples of the significant role of health research.

The vast amount of scientific knowledge accumulated to date, in addition to improvement in sanitation among other advances, particularly during the last decades, had made possible to reduce maternal and child deaths, decrease the prevalence of infectious diseases and gain better control of chronic diseases. These achievements have increased the life expectancy (LE) worldwide. Between 1950 and 2000, LE increased 3 years per decade, and from 2000 to 2015, it rose to 5 years per decade in most regions. The African Region shows the most significant improvement since LE increased 9.4 years. Furthermore, in the same period LE at age 60 also improved from 18.7 to 20.4 years at a global level [1]. However, this advancement is not equivalent to healthy life expectancy. In developing regions people live longer but not healthier. On average, there are 25 healthy life years lost due to morbidity and disability [1]. Chronic diseases rise with age, thus posing enormous challenges for health systems to respond to the growing health needs and demands of this age group. The top priorities for health systems are financial risk protection, universal access to health services, and quality of care. Fulfilling these three priorities is critical for older adults, since they are among the most vulnerable groups.

Health systems research (HSR) is a relatively novel area in health sciences that comprises the analysis of health systems policy, organization and provision at macro (e.g., country), meso (e.g., states) and micro (e.g., health facilities) levels. In addition, HSR furthers the development of innovative interventions and strategies to strengthen health systems and services for improving population health.

The Alliance for Health Policy and Systems Research define HSR "as a field that seeks to understand and improve how societies organize themselves in achieving collective health goals, and how different actors interact in the policy and implementation processes to contribute to policy outcomes. "By nature, it is interdisciplinary, it is a blend of economics, sociology, anthropology, political science, public health and epidemiology that together draw a comprehensive picture of how health systems respond and adapt to health policies, and how health policies can shape – and be shaped by – health systems and the broader determinants of health" [2]. This definition offers a comprehensive image of the significance of this research area in transferring new knowledge that enables health programs, interventions and services, to respond effectively to health population needs.

### 12.1.1 *Characteristics of Health Systems Research*

The primary goal of HSR is the strengthening of health systems to improve its performance and responsiveness to achieve better population health. One of the main HSR attributes is its applicability that translates into evidence at different levels that range from health policy makers to program managers and health care providers. Another core attribute of HSR is its capacity to generate context-specific data, with potential to expand to other scenarios. Both qualities depend on the researchers' ability for designing solid studies underpinned by sound conceptual and methodological aspects, to develop interventions adaptable to different contexts, and to scale-up these experiences as health programs and policies.

HSR is interdisciplinary; it encompasses a wide range of fields of knowledge that allows addressing the full range of health systems functions such as policymaking, governance/stewardship, financing, resource generation and provision of healthcare. In addition, it also addresses the outputs and outcomes of health systems and health services provision such as efficiency, effectiveness, equity, and quality of care.

HSR also embraces operational research, aimed at designing and testing strategies to solve specific health services provision problems. This area of research comprises the operational evaluation of programs or local interventions focusing on solving organizational problems [3]. Operational research requires special research abilities for overcoming barriers, such as inadequate perception of the research value that local health service managers and other stakeholders might have, and the capacity to identify structural or organizational obstacles that health services face to implementing health strategies and programs based on scientific evidence.

Implementation research is a novel field in HSR, in which new strategies are developed to set up interventions that have proven its efficacy and accessibility to be applied at health services level or scaled up at health system level, or even at multi-country scale, with the purpose of demonstrating its effectiveness. This type of research has particular frameworks to guide the design and methodological procedures, and the HSR community increasingly recognizes its relevance [4].

An essential characteristic of HSR is the participation of decision-makers in identifying and prioritizing the research question that often addresses the challenges to improving health system or services [4]. This feature is particularly relevant to develop feasible research projects within the routine operation of services, despite time constraints and health providers' resistance for being involved or be the subject of research studies (i.e., evaluations of quality of care).

Researchers should be aware of all HSR attributes mentioned earlier when answering questions aimed at improving health systems and services and of the methodological challenges. For example, the selection of the appropriate study design, study population, method of data collection, and operational and ethical aspects. The design of innovative interventions and impact evaluations of programs or services are complicated. Due to the natural resistance of the providers, researchers should offer convincing reasons on the relevance of the results of the proposed intervention for improving the efficiency and effectiveness of health services.

Systems thinking is a valuable approach for health systems analysis; it provides a set of tools useful to develop intervention and evaluation designs. Systems thinking can help "to explore problems from a system perspective; show potentials of solutions that work across sub-systems; promote dynamic networks of diverse stakeholders; inspire learning; and foster more system-wide planning, evaluation and research" [5].

### 12.1.2 Conceptual Frameworks for Health Systems Research Focused on Aging

The need of a conceptual framework for HSR is a relevant matter due to the complexity of this research area and the multiple dimensions and domains that its definition comprises. Several authors identified that such dimensions and domains have diverse meanings in low and middle-income countries than in high-income countries, and among different disciplines [6]. Therefore, HSR is not rooted in a unique conceptual framework, but in several frameworks; even though most of such frameworks contribute to understand the relevance of this research area.

The Health Systems framework that in the year 2000 the World Health Organization (WHO) defined, also encompasses HSR, which in turn underpins performance evaluations of health systems [7]. The WHO framework points out that improvement of health is the primary goal, and responsiveness to the needs and fair financial contribution of the people are two intermediate objectives for achieving health. Furthermore, health system performance depends on four core functions: (i) stewardship (oversight), (ii) generation of human resources (investment and training), and (iii) financing (collecting, pooling, and purchasing), which converge in (iv) service delivery [8]. Based on these concepts, HSR is appropriate to analyze the relationships between health system functions' performance on accomplishing better population health. However, the WHO framework has limitations to perform HSR on aging population, in which the analysis of health needs should consider a broader spectrum that comprises social determinants of health, coordination among public sectors (social networks, pensions, housing, and social services) to answer to the particular needs of the older adults that influence their wellbeing.

In 2008, the WHO published the World Health Report that emphasized Primary Health Care as the core of health systems [9]. This report underlines four broad essential concepts: universal coverage, people-centered care, integration of health into public policies across sectors, and health governance. The 2008 publication offset the limitations of 2000 report, since its conceptual framework reflected in a better way social values such as equity, response to people's needs and expectation, community participation and public-sector responsibility. Based on these principles, HSR has a comprehensive pathway to face the challenges of research on the aging population area.

Several other frameworks are available to understanding health systems performance, and different models can contribute to knowing HSR better. However, the vast diversity of research questions aiming at improving health systems performance in a demographic and epidemiological transition era imposes enormous challenges. Acknowledging that health care for population aging requires innovative models, HSR should continue searching fundamental questions and developing conceptual frameworks. As Hoffman et al., indicated: "The field must avoid being captured by any single paradigm, tradition or discipline, or excluding any perspective that may be important or helpful" [7].

#### 12.1.2.1 First Steps: The Research Question

Identifying the research problem is the critical challenge in HSR, and at the same time is the springboard for a valuable research project.

In addressing the HSR question, since the onset there are three sequential aspects useful for identifying the appropriate research topic (Table 12.1):

1. The health problem or health condition. Identification of health population needs is essential to define the magnitude of the problem to be addressed and the size of the population approach: national, subnational, communities, households, or individuals.
2. The health system or health services problem. Once the health condition is defined, the level of study should be determined: macro level (health system),

**Table 12.1** Sequence and examples to define with more precision the research question

| Sequence | Examples |
|---|---|
| The health problem | Magnitude: prevalence, incidence, mortality. |
| | Population affected, age, sex, ethnicity. |
| | Type of health condition: health risk, acute or chronic disease. |
| The health system/health services problem | At health system level: |
| | Population coverage |
| | Financial protection |
| | Resources (infrastructure or personnel) |
| | Governance/stewardship |
| | Quality |
| | At health services level: |
| | Access to facilities |
| | Services utilization: demand of services |
| | Affordability |
| | Quality of care |
| | Continuity of care |
| | Coordination of services |
| | Organizational aspects |
| The research problem | Health care: |
| | Needs |
| | Seeking |
| | Utilization |
| | Consequences |
| | Perspective: |
| | Health policy |
| | Health system |
| | Health services |
| | Population |

operational level (health services); additionally, the dimension where the problem is happening: access, health care provision, economic issues, etc.

3. The research problem. Characterization of the health condition and the problematic situation of the health system or health services facilitates identifying relevant research problems and the pertinent perspective for the study question.

Once the scientists identify the research problem, then, the formulation of a rational research question should consider the current knowledge of the selected topic, regarding:

a) The magnitude and distribution of the health condition
b) The causes or association with the health condition (risk factors)
c) Availability of effective interventions for tackling the health condition
d) Evaluation of the interventions implemented for improving the health condition

The literature review is a necessary condition to learn about the state of the art in scientific evidence and recognize the value of a research question, thus avoiding duplication. Replicating studies performed in other contexts and not previously carried out in the health services or health system of interest is valid but is essential to building a substantial justification of scientific relevance for the new research study. Such justification must support the additional knowledge that the research study will produce. Information about previous studies will also allow specifying the core question, regarding originality, relevance, and applicability of the study results. Also, it will guide to choose the type of study most suitable to respond the research question (Table 12.2).

Finally, formulating the research question should consider some challenges for the health systems researchers, which are particularly important to define the study [10, 11]:

1. Usefulness of the results. The applicability of the information that the study is expected to obtain is a key point for HSR. Therefore, the identification of

**Table 12.2** Purposes of research studies according to research questions

| Research question | Purpose of research study |
|---|---|
| What is the nature and extent of the research problem? | To explore little-understood situations at population or health systems/services. |
| | To describe the phenomenon. |
| What are the causes or consequences of the problem? | To identify the reasons for the problem |
| | To identify the associated factors with the problem. |
| What is the best solution for the problem? | Action research: To test strategies for improving health system/services performance or health status of the population. |
| How is the performance of health system, health services or health programs? | To evaluate: |
| | Policy or programs design. |
| | Process of care. |
| | Results of policies, strategies or programs. |
| | Impact of interventions. |

policy-relevant concerns of decision-makers is crucial to provide them with valuable results for being used for improving health systems or health services.

2. Feasibility. Research needs financial resources to be carried out and is the researcher' responsibility to obtain them; human resources, equipment, transportation are expenses required for HSR studies. Institutional or local capacity is seldom available and external financial support should be secured as a condition for a successful project.
3. Political acceptability. The involvement of decision-makers at the beginning of the process is fundamental to assure the interest and permissibility of the research topic approach. Conducting HSR requires the support of policy actors, convinced of the study relevance for improving their performance.
4. Ethical pertinence. The scope and context of the study should be considered for identifying cultural sensitivity of the population and providers. Acceptability through informed consent is mandatory, and sharing study results to participants at the end of the project must be warranted.

## 12.2 Study Designs and Methods in Health Systems Research

Health systems research aims at building scientific evidence to inform stakeholders (e.g., decision-makers, healthcare providers, citizens and civil society) on the population health needs and performance of the health systems, as well as on scientifically sound policies, programs, and interventions for improvement [12]. To respond to the complex demands of the health systems, HSR incorporates multi-disciplinary research, combining multiple research designs and diverse methods of data collection and analysis.

As in other disciplines, the design of HSR studies should be guided by the research question (e.g., exploratory, descriptive, or causal), data sources (e.g., primary or secondary; qualitative or quantitative) and methods for data collection (e.g., literature review, focus groups, direct field observations, surveys, administrative or clinical records, etc.) and analysis (e.g., descriptive, inferential) [13].

Previous chapters of this book describe several HSR designs and methods (e.g., systematic review and meta-analysis, qualitative research, descriptive studies, longitudinal research, clinical trials, mixed-methods research). Therefore, we will only focus on those designs and methods not mentioned before or need further clarifications. Mainly, we will briefly describe cross-sectional studies, case studies, implementation research, impact and economic evaluations in HSR.

### 12.2.1 Cross-Sectional Studies

Cross-sectional design focuses on analyzing data collected in one point in time [13]. It can be conducted through a survey or analysis of existing databases (e.g., administrative, or electronic health record databases). This design is useful for different purposes (e.g., descriptive, analytical). Researchers use this design very often, since it saves time and money, although its main limitation is that cannot be used to making causal inference. In HSR focused on older adults, cross sectional studies are used often to evaluate their health related needs (e.g., supportive care needs for activities of daily living, psychological problems, information needs etc.). In addition, this design is useful to analyze some aspects of the quality of healthcare that older adults receive, and to investigate factors associated with diverse health problems of older adults that allow researchers to generate hypothesis that eventually longitudinal studies could test.

As an example, in a study conducted in Mexico, the researchers analyzed the 2006 Survey of Autonomy and Dependency (EDAD–IMSS; N = 3348) that aimed at identifying the different types of the social networks (TSNs) of older adults affiliated with the Mexican Institute of Social Security. The results described the main characteristics of older adults in each TSN, including the instrumental and economic support they received; also, the study served to determine the association between functional dependency and the TSN. Researchers identified the TSNs through using a structural approach and cluster analysis. The association between functional dependency and the TSN was evaluated with Poisson regression with robust variance analysis in which socio-demographic characteristics, lifestyle and medical history covariates were included. This study found five TSNs of older adults: diverse with community participation (12.1%), diverse without community participation (44.3%); widowed (32.0%); non-friends-restricted (7.6%); nonfamily-restricted (4.0%). Older adults belonging to widowed and restricted networks showed a higher proportion of dependency, negative self-rated health and depression [14].

Another example is a descriptive secondary data analysis of patients' health records to evaluate the quality of medication prescribing. The study analyzed records of 495 subjects aged 60 years of age or older, who were seen due to non-malignant pain syndrome in primary care clinics. The Beers criteria and the medication appropriateness index (MAI) were the tools to assess the quality of medication prescribing. The study found that about 35% of patients had prescriptions with at least one inappropriate medication. According to MAI the most frequent errors were failure to provide practical instructions, prescription of drug combinations with potential drug-drug interactions, and inappropriate indications [15].

### 12.2.2 Case Studies

In HSR, case studies focus on complex HS related phenomenon within real-life contexts (e.g., a specific program in a single facility such as health centers, hospitals, or community mental health centers; or a particular health legislation in a specific country) [16, 17]. Case studies design can be divided into two groups: single and multiple (comparative) case studies. Yin identified several important attributes of HSR case studies: (1) research question(s) should focus on "how" and "why." (2) The theoretical proposition should guide operational definition of studied phenomenon, unit of analysis, selection of case(s), data collection, analysis, and interpretations. (3) A single case study can serve for theory-related analytic generalization; while, a comparative case study design allows external generalization by using replications, especially when there is a purposeful selection of cases. (4) Rival explanations should be defined and tested; (5) Data collection should involve triangulation of evidence from multiple sources (qualitative and quantitative data sources).

The examples of single case studies in HSR focusing on the needs of the elderly describe specific health systems policies that took place in Australia [18] and Chile [19]. An example of a comparative case study is a research on the nature and functioning of eight different public policy networks in the United Kingdom National Health Service, which included regional and metropolitan Older People networks [20]. The last study was based on theoretically informed questions obtained through initial literature review and used triangulation of information through different data sources, such as documents, notes taken at attendance meetings and semi-structured interviews with a range of key stakeholders. Examples for case studies referred to individuals and not health systems can be found in Chap. 5.

### 12.2.3 Implementation Research

Implementation research seeks to understand barriers to and facilitators of implementation of policies, programs, or individual interventions into routine use, in usual care settings to provide guidance on their systematic uptake [21–23]. Implementation research is a final step in the traditional translational research continuum after efficacy and effectiveness studies [24]. In contrast to the randomized efficacy and effectiveness trials that focus on internal validity, implementation research puts emphasis on external validity [24]. It plays significant attention to the context in which interventions are adopted, scaled up, and sustained. According to the WHO, "the context can include the social, cultural, economic, political, legal, and physical environment, as well as the health systems and institutional settings, comprising various stakeholders and their interactions, and the demographic and epidemiological conditions" [23].

To accelerate the adoption of the intervention(s) the simultaneous evaluation of effectiveness and implementation aspects has been proposed by using effectiveness-implementation hybrid designs [25]. The studies with these designs apply broad eligibility criteria, approximating "real world" use. In addition, hybrid and pure implementation studies employ a broad variety of qualitative, quantitative, and mixed methods. One of the examples, of this type of research, is a study on the implementation of an Evidence-Based Exercise Program for Older Adults in South Florida [26]. This study included 14 community-based agencies in 83 different locations and reached 4490 older adults. The study combined effectiveness and implementation evaluation using multiple sources of information, including descriptive/demographic data form, a health history register, a fitness check, and program satisfaction survey. Another example is a report on the implementation of the Whiringa Ora [27] community-based program in New Zealand aimed at facilitating interdisciplinary care for rural patients with a chronic disease who have high inpatient admissions or emergency department presentations. The study focused on evaluating the barriers, facilitators and impact of the program, and included a mixed-method design that combined qualitative and quantitative methods, such as in-depth interviews with patients and healthcare providers, analysis of patient records and telehealth measures on clinical outcomes and hospital resource utilization and satisfaction surveys.

### 12.2.4 Impact Evaluation

*"Impact evaluation (IE) is a particular type of evaluations that seeks to answer cause-and-effect questions about the impact (causal effect) of a program on an outcome of interest"* [28]. IE is an important mechanism that serves to identify the effect of public policies and programs. The results of well-implemented IEs can provide accurate and reliable information for accountability and evidence-based decision-making purposes [28]. Yet, these results do not provide insights into program implementation.

IE can be performed prospectively and retrospectively. Contrary to the prospective evaluation, retrospective IE as any secondary data analysis may lack data on essential outcome variables and covariates that will limit the extent of the study conclusions.

Estimating causal effects in IE centers on the counterfactual that is *"an estimate of what the outcome (Y) would have been for a program participant in the absence of the program (P)"* [28]. At the same time, a same person cannot participate in a program and comparison groups; therefore, it is necessary to identify a valid comparison group with similar characteristics to the program participants. To obtain a valid comparison, both groups: (1) must have on average the same observed and unobserved characteristics; (2) must react to the program in the same way; (3) cannot be differentially exposed to other interventions during the evaluation period. *"When these three conditions are met, then only the existence of the program (P) of interest will explain any differences in the outcome (Y)"* between intervention and control groups [28].

The methods to produce valid comparison groups include: I. Methods that generate estimates of the counterfactual through explicit program assignment rules, based on the:

a) Randomized assignment of the treatment, comprising: i) Randomized offering of the treatment – when researchers randomly select participants to whom they offer the treatment; yet, usually not everyone complies with the assignment. ii) Randomized promotion of the treatment or encouragement design, where the program is open to everyone and the researchers randomly select participants to whom they will promote the program. This method seeks to increase the uptake of a voluntary program in a subsample of the population.
b) Regression Discontinuity Design for programs with a continuous eligibility index with a defined cutoff score (e.g., pension program, where only people above a certain age are eligible to receive the pension). This design estimates the average impacts around the eligibility cutoff at the point where treatment and comparison participants are most similar.

II. When randomization or regression discontinuity designs are not feasible, observing only the before-and-after change in the program outcomes will not accomplish estimating the program's causal effect because other factors are likely to influence on reasons for enrollment and on the outcomes. To avoid these possible selection bias, two other methods were proposed [28]:

a) Difference-in-differences (DD) method *"estimates the counterfactual for the change in outcome for the treatment group by calculating the change in outcome for the comparison group"* [28]. The DD estimation include calculation of the difference in the outcome between the before and after data for the treatment (B − A) and comparison (D − C) groups and calculation of the difference between the difference in outcomes for the treatment and the comparison groups (B − A) − (D − C). This method allows taking into account any differences between the treatment and comparison groups that are constant over time. However, it does not help eliminating the between-groups differences that change over time. Therefore, we must assure (or assume) that there are no such time-varying differences between the groups. If this assumption is not met and there is a factor that affects only the treatment group, when it receives the treatment; consequently, the DD estimate might be biased.
b) Propensity score matching is *"a set of statistical techniques used to construct the best possible comparison group for a given treatment group based on "propensity score" (PS)* [28]. PS represents the probability (from 0 to 1) of participant's enrollment to the program derived from the values of their observed characteristics. PS is used to match participants in the treatment group with those in the comparison group based on the closest PS. Matching is a quasi-experimental approach which attempts to replicate randomized assignment to ensure that all determinants of outcomes (other than treatment status) are similar between the treated group and their matched comparisons. The necessary assumptions for PS matching are (a) conditional independence and (b) the presence of a common

support (overlap between propensity score distributions of treatment and comparison groups). The first assumption means that after controlling for observable covariates, the potential outcomes are independent of treatment status and unobserved characteristics that could affect the outcome exist between the treatment and the matched comparison group. Therefore, when PS matching reduces overt selection bias owing to observed differences among study groups, it does not permit controlling for hidden unobserved differences.

In addition, the higher robustness of the estimated counterfactual can be achieved by combining different impact evaluation methods (e.g., matched difference-in-differences or difference-in-differences regression discontinuity design).

Impact evaluations are highly used in HSR of policies and programs focused on the aging population. For example, in Italy, Battistin et al. [29], used regression discontinuity approach to investigate the extent of the consumption decrease at retirement in Italy. The researchers assumed that consumption would be the same around the threshold for pension eligibility if individuals would not retire. The study showed that the fall in consumption could be explained by drops in work-related expenses, leisure substitutes and the number of grown children living with their parents.

In Taiwan, Liao et al. [30], used longitudinal data from the Survey of Health Living Status of the Elderly and a difference-in-differences method to investigate the effects of the introduction of Taiwan's National Health Insurance (NHI) on life satisfaction within the elderly. The study found that compared to the change in life satisfaction between the previously uninsured and insured senior men, the introduction of NHI had a more significant effect of 4.33 points on reducing the disparity in life satisfaction between previously uninsured and insured elderly women.

In Mexico, Doubova et al. [31], used data from the 2012 Survey of Health and Nutrition (ENSANUT) and PS matching technique to analyze the effects of Social Security (SS) and Seguro Popular (SP) health insurance (HI) on access to healthcare of older adults, and on the financial risk protection to their households in comparison with the population without health insurance. After matching on observable characteristics, researchers found that SP was inferior to SS-HI both in providing access to healthcare and in assuring financial risk protection for Mexican older adults.

In China, Wang et al. [32], used longitudinal data from the household surveys conducted one-year pre-, and 2 years post-intervention and DD estimate combined PS matching method to evaluate the impact of community-based health insurance scheme called "Rural Mutual Health Care (RMHC)" on enrollees' health status measured by EQ-5D. The results show that RMHC has a positive effect on the health status of participants, significantly reducing pain/discomfort and anxiety/depression for the general population and has a positive impact on mobility for those over 55-years old.

### 12.2.5 Economic Evaluation

Economic evaluation is a comparative analysis of alternative courses of action (e.g., policies, programs, specific interventions) regarding costs and consequences [33]. The primary reason for undertaking an economic evaluation is to produce information on the efficient use of the resources in the healthcare sector. In fact, the economic evaluation is an essential tool for priority setting [34].

Economic evaluation covers full economic evaluation studies (e.g., cost-benefit analysis, cost-utility analysis, cost-effectiveness analysis) and partial economic evaluation studies that consider costs and/or consequences, but which either do not involve a comparison between alternative interventions or do not relate costs to the benefits (e.g., cost analyses, cost-description studies and cost-outcome descriptions) [35]. The decision on the economic evaluation method requires considering potential uses for resource-allocation decision-making, availability of information and computational complexity [36].

Economic evaluations can be conducted from different perspectives that determine costs and consequences included in the analysis. There are providers, public-payer, patient, or societal perspectives. The most comprehensive is a societal perspective that captures the value of all changes in resources used and gained as a result of an intervention, including informal caregiving and productivity costs [37].

Economic evaluations of health interventions can be based on data from a specific randomized controlled trial (RCT) or use decision analytical modeling [33]. A single randomized trial has some limitations for comprehensive economic evaluation; Such as a short follow-up period inadequate to capture differences in important outcomes. Other limitations are the use of intermediate measures instead of the final health outcomes, narrow inclusion criteria, and specific local characteristics, that reduce the scope of the study's conclusions [38, 39]. Conversely, decision analytical modeling use evidence from different trials, meta-analyses, and observational studies that allow estimations based on aggregate data from different sources, which consequently avoid single trial limitations [39]. The two frequently used model types are decision trees and Markov models.

The most frequently used in HST is a cost-effectiveness analysis (CEA). CEA compares the relative costs and outcomes (effects) of alternative interventions/programs to select the one with the highest effectiveness relative to its cost. Different effectiveness measures can be used in CEA, such as process measures (e.g., professional guidance adherence, patient adherence to medication), or health outcomes (e.g., mortality, or new cases of disease that were prevented, decrease in intensity of the disease symptoms, etc.) [35, 36]. The summary measure of CEA is a cost-effectiveness ratio (e.g., cost per case averted, cost per life-year saved) at patient or population level. Incremental cost-effectiveness ratios are established by dividing the difference in costs of various implementation strategies by the corresponding difference in health outcomes [36].

*Cost–Benefit Analysis (CBA)* is a full economic evaluation that estimates the total expected benefits of a program, compared to its total expected costs. CBA assigns a monetary value to both costs and benefits of alternative interventions, assessing whether benefits of different interventions outweigh their costs [28, 35, 36].

*Cost–Utility Analysis (CUA)* compares alternative interventions regarding cost per unit of utility gained. Utilities are measures that comprise both life expectancy and subjective levels of well-being of each intervention. The best-known utility measure is the quality-adjusted life year (QALY). The number of QALYs derived from a health intervention is the lifetime gained because of it, weighted by the health-related quality of life (HRQoL). However, the use of QALYs in economic evaluation of interventions that target older adults has several disadvantages [37, 40]. These include: (1) The reduced life expectancy of older adults usually reflects in lower gains in QALYs "produced" by an intervention for elderly population, compared to the intervention for younger people; accordingly, the intervention for older people will be less valued. (2) The health gains that can be obtained in older adults are smaller, even if an intervention is successful it will not restore full health. At the same time, small health gains are measured poorly by instruments used to assess QALYs. (3) QALYs consider HRQoL, ignoring other effects of interventions important for older adults like social effects; (4) EQ-5D and other instruments measure HRQoL regardless of age. "The HRQoL of older people will be underestimated if age-dependent measures of value are not considered" [40].

Despite a large number of interventions aimed at maintaining and improving the health of older adults, only a few assess its cost-effectiveness. Most CEA studies come from high-income countries; however, this type of studies is of paramount importance in low and middle-income countries due to the scarcity of resources for health.

For example, a systematic review of the literature carried out by Tappenden et al. [41], identified only three cost-effectiveness studies of home-based, nurse-led health promotion for older people. This report concluded that nursing care at home has a lower cost compared to hospital care; however, the authors acknowledged that due to the scarce scientific literature, the evidence of the cost-effectiveness of this type of services for health systems is still lacking.

In the US, Gitlin and colleagues conducted a cost-effectiveness analysis of an Tailored Activity Program (TAP) for individuals with dementia and family caregivers, which consisted of eight sessions by occupational therapists over 4 months, comparing it with home care (conventional care) [42]. Over a 4 month time horizon, the group of patients with the intervention showed a significant improvement in functionality and lowered the burden on the caregiver compared to the group of patients with conventional care. The cost of the intervention was $ 942 US dollars. The results of the Monte Carlo model showed that the intervention was cost-effective about 80% of the time.

The study conducted in the Netherlands, Graff et al. [43], performed a cost-effectiveness analysis of an intervention that consisted of 10 sessions of occupational therapy in patients with dementia compared to conventional care without occupational therapy. In a time horizon of 5 weeks, it was found that the cost of the intervention was $ 1738 US dollars; being cost-effective for the improvement of the patients and in the decrease of the burden of the caregiver.

## 12.3 Overall Recommendations

Researchers working in HSR arena should be able to understand the importance and the scope of performing a type of research that has multiple and different layers that range from the individual to the complicated international scenario. HSR is a cross-pollination of knowledge coming from different disciplines. To be able to perform this type of research, scientists must be able to work collaboratively with peers that have diverse conceptual backgrounds and scientific philosophies. Furthermore, they must develop the capacity to listen and raise the interest of various stakeholders such as donors, politicians, and the industry, among others. Also, researchers should acquire the skills to keep a policy dialogue, learn about the context of the health systems, identify the problem and the research question, work closely with decision makers, providers, and the public, and be able to translate the scientific evidence into persuasive messages and recommendations for all parties involved. The interaction between researchers and decision-makers should be constructive; though researchers must understand that political, economic, social and personal agendas influence the decision-making process. The meaningful scientific evidence is another valuable component of such process.

HSR influences and speeds up the evolution of health systems and the body of knowledge is building up through assembling experimental data and theoretical concepts. Health systems are dynamic and evolve quickly, yet in many countries, the political and financial contexts shape the architecture of the health systems and the health policy agenda. Given the wide variety of arrangements and evolutionary stage of health systems globally, along with the complex mix of unsolved and emerging health needs and demographic changes, it seems that HSR is a Sisyphean task. As in other research fields, conducting HSR means perspiration rather than only inspiration and the effort should be tangible with the evidence that performance of health systems and health status of the population are improving. Thus, HSR scientists must try harder in their efforts to make better contributions.

## 12.4 Conclusions

The contribution of HSR to improve health systems and health services for the aging is particularly relevant. We all are aware that the elderly has specific health needs, and in developing countries, their health status is more vulnerable, given the social and economic conditions they endure throughout life, thus increasing the probability of chronic illnesses and dependency. The aging of the population, the rising trends of dependency and the poor health status of the older adults, impose important challenges for any health system. Thus, the contribution of HSR should be able to strengthen the capacity to respond to such challenges.

The chapter provides a general description of study designs and methods in health systems research focused on aging population; it is intended to be descriptive and able to give an appropriate springboard for those who plan to do HSR. We are aware that are excellent textbooks for each of these designs. We directed our effort to provide a comprehensive view and practical examples from our experience and the literature.

## References

1. World Health Organization (2016) Chapter 3: Monitoring the health goal – indicators of overall progress. In: World health statistics 2016: monitoring health for the SDGs, sustainable development goals. World Health Organization, Geneva http://www.who.int/gho/publications/world_health_statistics/2016/EN_WHS2016_Chapter3.pdf?ua=1. Accessed 16 Jan 2018
2. Alliance for Health Policy and Health Systems Research (2011) What is HPSR? Overview. World Health Organization, Geneva http://www.who.int/alliance-hpsr/about/hpsr/en/index.html
3. Remme JHF, Adam T, Becerra-Posada F, D'Arcangues C, Devlin M, Gardner C et al (2010) Defining research to improve health systems. PLoS Med 7(11):e1001000. https://doi.org/10.1371/journal.pmed.1001000
4. Kirk MA, Kelley C, Yankey N, Birken SA, Abadie B, Damschroder A (2016) A systematic review of the use of the consolidated framework for implementation research. Implement Sci 11:72. https://doi.org/10.1186/s13012-016-0437-z
5. De Savigny D, Adam T (2009) Systems thinking for health systems strengthening. World Health Organization, Alliance for Health Policy and Systems Research, Geneva
6. Hoffman SJ, Røttingen JA, Bennett S, Lavis JN, Edge JS, Frenk J (2012) Conceptual issues related to health systems research. Working paper. Alliance for Alliance for health policy and systems research. World Health Organization, Geneva
7. Murray CJL, Frenk J (2000) A framework for assessing the performance of health systems. Bull WHO 78(6):717–731
8. World Health Organization (2000) The World health report 2000: health systems: improving performance. World Health Organization, Geneva
9. World Health Organization (2008) The world health report 2008: primary health care – now more than ever. World Health Organization, Geneva
10. Varkevisser CM (1993) Designing and conducting health systems research projects: volume 1: Proposal development and fieldwork. IDRC
11. Gilson L (ed) (2013) Health policy and systems research: a methodology reader. Alliance for health policy and systems research. World Health Organization, Geneva
12. WHO (2012) Strategy on health policy and systems research: changing mindsets. World Health Organization, Geneva
13. Shi L (1997) Health services research methods. Delmar Publishers, Albany
14. Doubova SV, Pérez-Cuevas R, Espinosa-Alarcón P, Flores-Hernández S (2010) Social network types and functional dependency in older adults in Mexico. BMC Public Health 10:104
15. Doubova SV, Torres-Arreola LP, Rosas-Carrasco O, Pérez-Cuevas R (2005) Calidad de la prescripción en los adultos mayores co síndrome doloroso de origen no oncológico usuarios de los servicios de Medicina Familiar. Rev Investig Clin 59(6):428–436
16. Yin RK (1994) Case study research design and methods, 2nd edn. Sage, Thousand Oaks
17. Yin RK (1999) Enhancing the quality of case studies in health services research. Health Serv Res 34(5 Pt 2):1209–1224

18. McPake B, Mahal A (2017) Addressing the needs of an aging population in the health system: the Australian case. Health Syst Reform 3(3):236–247
19. Thumala D, Kennedy BK, Calvo E, Gonzalez-Billault C, Zitko P, Lillo P et al (2017) Aging and health policies in Chile: new agendas for research. Health Syst Reform 3(4):253–260
20. Ferlie E, Fitzgerald L, Mcgivern G, Dopson S, Bennett C (2011) Public policy networks and 'wicked problems': a nascent solution? Public Adm 89(2):307–324
21. Peters DH, Adam T, Alonge O, Agyepong IA, Tran N (2013) Implementation research: what it is and how to do it. BMJ 347:f6753. https://doi.org/10.1136/bmj.f6753
22. Roundtable on Translating Genomic-Based Research for Health Board on Health Sciences Policy; Health and Medicine Division; National Academies of Sciences, Engineering, and Medicine. Applying an implementation science approach to genomic medicine: workshop summary. National Academies Press, 2016. Available from: https://www.ncbi.nlm.nih.gov/books/NBK373700/
23. Peters DH, Tran NT, Adam T (2013) Implementation research in health: a practical guide. Alliance for Health Policy and Systems Research, World Health Organization, Geneva
24. Brown CH, Curran G, Palinkas LA, Aarons GA, Wells KB, Jones L et al (2017) An overview of research and evaluation designs for dissemination and implementation. Annu Rev Public Health 38:1–22
25. Curran GM, Bauer M, Mittman B, Pyne JM, Stetler C (2012) Effectiveness-implementation hybrid designs: combining elements of clinical effectiveness and implementation research to enhance public health impact. Med Care 50(3):217–226
26. Palmer RC, Batra A, Anderson C, Page T, Vieira E, Seff L (2016) Implementation of an evidence based exercise program for older adults in South Florida. J Aging Res 2016:9630241
27. Carswell P (2015) Te Whiringa Ora: person-centred and integrated care in the Eastern Bay of Plenty, New Zealand. Int J Integr Care 23(15):e014
28. Gertler PJ, Martinez S, Premand P, Rawlings LB, CMJ V (2011) Impact Evaluation in Practice. The International Bank for Reconstruction and Development/The World Bank, Washington DC Available at http://www.worldbank.org/pdt
29. Battistin E, Brugiavini A, Rettore E, Weber G (2008) The retirement consumption puzzle: evidence from a regression discontinuity approach, IFS working papers, No. 08,05. Available at https://www.econstor.eu/bitstream/10419/47479/1/578421895.pdf
30. Liao PA, Chang HH, Sun LC (2012) National health insurance program and life satisfaction of the elderly. Aging Ment Health 16(8):983–992
31. Doubova SV, Pérez-Cuevas R, Canning D, Reich MR (2015) Access to healthcare and financial risk protection for older adults in Mexico: secondary data analysis of a national survey. BMJ Open 5:e007877. https://doi.org/10.1136/bmjopen-2015-007877
32. Wang H, Yip W, Zhang L, Hsiao WC (2009) The impact of rural mutual health care on health status: evaluation of a social experiment in rural China. Health Econ 18(Suppl 2):S65–S82
33. Drummond MF, O'Brien B, Stoddart GL, Torrance GW (1997) Methods for the economic evaluation of health care programmes, 2nd edn. Oxford University Press, Oxford
34. Hauck K, Smith PC, Goddard M (2004) The economics of priority setting for health care: a literature review. Health, Nutrition and Population (HNP) Discussion Paper. The International Bank for Reconstruction and Development/The World Bank, Washington, DC
35. Economics and economic evaluation (n.d.). In: Higgins JPT, Green S. Cochrane handbook for systematic reviews of interventions. Version 5.1.0. [updated March 2011]. Available at: http://handbook-5-1.cochrane.org/chapter_15/15_1_2_economics_and_economic_evaluation.htm
36. Hoomans T, Severens JL (2014) Economic evaluation of implementation strategies in health care. Implement Sci 9:168. https://doi.org/10.1186/s13012-014-0168-y
37. Huter K, Kocot E, Kissimova-Skarbek K, Dubas-Jakóbczyk K, Rothgang H (2016) Economic evaluation of health promotion for older people-methodological problems and challenges. BMC Health Serv Res 16(Suppl 5):328. https://doi.org/10.1186/s12913-016-1519-y

38. Sculpher MJ, Pang FS, Manca A, Drummond MF, Golder S, Urdahl H et al (2004) Generalisability in economic evaluation studies in healthcare: a review and case studies. Health Technol Assess 8(49):iii–iiv 1–192
39. Petrou S, Gray A (2011) Economic evaluation using decision analytical modelling: design, conduct, analysis, and reporting. BMJ 342:d1766. https://doi.org/10.1136/bmj.d1766
40. Edlin R (2014) Assessing the cost-effectiveness of therapies for older people. In: Harper S, Hamblin K (eds) International handbook on ageing and public policy. Edward Elgar, Cheltenham/Northampton, pp 167–177
41. Tappenden P, Campbell F, Rawdin A (2012) The clinical effectiveness and cost-effectiveness of home-based, nurse-led health promotion for older people: a systematic review. Health Technol Assess 16(20):1–72
42. Gitlin LN, Hodgson N, Jutkowitz E et al (2010) The cost-effectiveness of a nonpharmacologic intervention for individuals with dementia and family caregivers: the tailored activity program. Am J Geriatr Psychiatry 18:510–519. https://doi.org/10.1097/JGP.0b013e3181c37d13
43. Graff MJ, Adang EM, Vernooij-Dassen MJ et al (2008) Community occupational therapy for older patients with dementia and their care givers: cost effectiveness study. BMJ 336:134–138. https://doi.org/10.1136/bmj.39408.481898.BE

# Chapter 13
# Technology and Aging: Ubiquitous Sensing Technology for Aging Research

**Jesús Favela and Luis A. Castro**

**Abstract** Advances in Information and Communication Technologies (ICT) are impacting aging research in multiple ways, ranging from analyzing large volumes of data from longitudinal studies to assessing the efficacy of assistive robots. This chapter focuses on using ubiquitous technologies for gathering behavioral data from individuals to understand how we age, assess the effectiveness of interventions, perform early diagnosis of diseases, or monitor disease progression. The ubiquity of inexpensive sensors, most notably in mobile and wearable devices, and advances in pattern recognition algorithms capable of reliably inferring activities and behavior is providing a new and powerful tool for aging research. We describe how these technologies can be used to monitor clinical variables and health outcomes in interventions for aging and illustrate their use with case studies on assessing frailty, inferring anxiety in caregivers of people with dementia and monitoring eating behaviors. We conclude by discussing some of the issues facing research in this area regarding data quality and privacy.

**Keywords** Gerontechnology · Ubiquitous sensing · ICT · Technology and aging

## 13.1 Introduction

Advances in ICTs are impacting aging research in numerous ways. One notable case is the use of sensing technologies to support epidemiological studies and clinical interventions in aging. Behavioral epidemiology studies how lifestyle and behavior relate to the occurrence of a disease and evaluates interventions aimed at

J. Favela (✉)
Computer Science Department, Center for Scientific Research and Higher Education of Ensenada, Ensenada, Baja California, Mexico
e-mail: favela@cicese.mx

L. A. Castro
Department of Computing and Design, Sonora Institute of Technology (ITSON), Ciudad Obregon, Sonora, Mexico
e-mail: luis.castro@acm.org



C. García-Peña et al. (eds.), *Aging Research - Methodological Issues*,
https://doi.org/10.1007/978-3-319-95387-8_13

changing unhealthy behaviors, such as overeating or smoking. Numerous diseases provide early evidence of their onset from changes in behavior, long before confirmed by clinical studies. In particular, mobile phones and wearable devices include a variety of sensors that can be used to gather data about users' behavior, such as the places they visit, their level of activity, and how frequently and with whom they socialize. The collection and analysis of these data have been the focus of recent attention in an emerging field known as mobile sensing, which can offer valuable data to aging research. One such example is the Health eHeart study at the University of California at San Francisco, which used smartwatches to collect heart rate data from 196,000 participants that have been used to predict hypertension, sleep apnea, and diabetes with 80 to 90% accuracy [1]. Epidemiological studies, such as the Framingham Heart study and the Women's Health study at Harvard, are increasingly incorporating data obtained from mobile and wearable sensors. A notable example is the Precision Medicine Initiative in the U.S., which aims at creating a cohort of one million participants who will contribute biological and genetic data, as well as behavioral and lifestyle information derived from sensors in the mobile devices they use and wear.

This chapter presents three case studies that exemplify how sensing technology can be used to support research on aging. Section 13.2 describes the field of mobile sensing for healthcare, including the ubiquity of mobile and environmental sensors, and how data on human behavior can be derived from it. We then present three case studies that illustrate how data for aging research can be obtained from individuals and populations. Finally, in Sect. 13.3 we discuss some of the challenges facing the development of this field regarding data quality and addressing privacy concerns.

## 13.2 Activity and Behavior Monitoring Through Mobile and Environmental Sensing

Self-report is a technique commonly used to collect data for aging research using questionnaires or surveys. Through this method, data regarding behaviors, activities, and beliefs are collected directly from the individual, or a third person such as family member or caregiver. While convenient and useful, self-report can provide unreliable information from users who might not be sufficiently aware of their activities, might not remember important details of their behavior, might be biased, inclined to exaggerate, or simply lie when answering questions from the researcher.

In contrast with self-report, studies that rely on Real World Evidence (RWE) use information obtained from data gathered from patients in real life settings. This is increasingly made possible by the ubiquity of mobile and wearable devices capable of sensing information related to the user's environment and the development of algorithms capable of inferring the activity and behavior of the users of these devices. For instance, while a frailty questionnaire might include questions on the amount of physical activity performed by the patient in the last few days (i.e. frequency of performing moderate activity, such as walking), data from an accelerometer worn by the patient can be used to give a fair estimate of the number of steps walked, overall physical activity, and number of calories burnt.

Behaviors, mannerisms, or actions that can be estimated with good accuracy using mobile and wearable sensors and are relevant to research on aging, include physical activity, sleep, coughing [2], anxiety [3], or socialization [4]. These advances are making possible the use of behavioral cues as the basis for novel computing systems that can be used to support a range of healthcare solutions. This includes early diagnosis on conditions that might have distinct behavioral symptoms, such as Parkinson's disease; ambulatory assessment of patients under observation, such as those who have undergone surgery; assisting in the management of a disease by for instance recommending medication doses according to how it affects behavior in that individual; detecting problematic behaviors in conditions such as dementia to intervene before such behavior harms the patient (e.g., wandering); and in interventions aimed at inducing behavior change to assist the user to adopt healthier behaviors such as exercising more or stop smoking.

Mobile sensing is mainly carried out with modern mobile phones and wearables such as smartwatches. Those devices have been augmented with several built-in sensors that can provide information relevant to health studies. Some of these devices include up to 15 hardware-based sensors such as accelerometer, gyroscope, magnetometer, GPS, microphones, proximity, luminosity, among others. These sensors can be used to derive certain user behaviors. For instance, the accelerometer can be used for estimating gait speed, step counting, or physical activity. There are several technological frameworks that can help rapidly deploy these types of studies such as Funf (http://www.funf.org), AWARE [5] or InCense [6]. Deriving relevant healthcare variables from sensed data could be relatively straightforward, such as estimating geographic lifespace from GPS traces. However, other variables may require more sophisticated pattern recognition algorithms such as those used to estimate gait speed from accelerometer data from a smartphone, or even novel deep learning algorithms that have been used to find promising correlations between heart rate and diabetes. Interdisciplinary work from healthcare specialists and data scientists in this field will redefine how healthcare data are captured, analyzed, and used to diagnose, assess and manage healthcare.

In this section we illustrate how mobile and environmental sensing technology can be used to more reliably measure parameters of interest for functional assessment, detecting anxiety in caregivers and monitoring eating habits, which are traditionally obtained from self-report. Using sensing for assessment enables data to be gathered opportunistically as informants perform everyday activities through unobtrusive and ubiquitous sensors, such as mobile phones. This also allows for continuous monitoring rather than requiring patients to attend a clinic to complete surveys or be interviewed. We first present how mobile phones can be used to obtain behavioral data related to frailty, including data on mobility and activity. We then describe how wearable sensors can be used to detect anxiety in caregivers, and, finally, we describe how mobile technology can be used to monitor eating habits.

### 13.2.1 *Functional Assessment of Older Adults Using Smartphones*

Functional assessment often relies on self-report or is carried out at the doctors' office. This approach has validity problems as patients may under-report or exaggerate symptoms. Also, patient assessment at the clinic has ecological validity problems since it is based on occasional physical tests performed in a laboratory setting.

One such test at the doctor's office may include the Timed Up and Go test, which is used to assess older adults' functional mobility, gait speed, and risk of falling. Frenken et al. at OFFIS in Germany proposed an unsupervised approach to perform an equivalent test using ambient sensors in a domestic environment [7]. The test can be performed continuously, given the physician a more reliable assessment of functional mobility than a test that is performed in a lab every few months, at best. Walking speed and fatigue are among the factors associated with the frailty syndrome [8]. Unobtrusively monitoring gait speed over a period of time, for instance in a route frequently walked by an individual, could provide early evidence of fatigue.

The frailty syndrome is of particular interest for the functional assessment of older adults Frailty is a state of increased vulnerability to adverse health outcomes for people of the same age [9]. The frailty syndrome involves several aspects such as involuntary weight loss, exhaustion, muscle weakness, slow walking speed, and low physical activity [8]. Frail people are at high risk for major adverse health outcomes, including disability, falls, institutionalization, hospitalization, and mortality [10]. The clinical assessment of older adults is to a large extent based on retrospective accounts of incidents. This can be unreliable as patients often do not remember or try to hide or minimize negative incidents. For example, widely-accepted instruments to estimate frailty in older patients include questions such as "In the last week, in how many days you walked at least 10 min?" and "How frequently do you speak with your friends/spouse?" Often, responses to these questions are hardly precise, having older adults providing rather vague answers to questions of this nature. Therefore, mobile sensing represents an attractive approach for estimating some of those variables (e.g., physical activity, socializing with others) that could correlate to surveyed data of older adults pertaining to frailty.

We used a mobile phone and InCense [6] to assess frailty in older adults and compare it with the results of the clinical assessment. We recruited 15 community-dwelling older adults, average age was 75.3 (SD = 1.8), for gathering data from their activities and behaviors. Four of our participants were classified as frail, based on the frailty index developed by Fried [8]. We collected data from several sensors in the mobile phones including location, audio, and others. The functional assessment of frailty in elders included standard inventories such as the Katz instrument for measuring activities of daily living [11] and the SF-36 health survey [12].

From our results, which can be consulted at [6], we found that that sleeping time in frail participants was larger than for fit participants. Also, not surprisingly, frail participants went out of their households less often than fit participants. In addition, we found that participants who were fit performed significantly more intense activity bouts than those who were frail. Finally, we were also able to estimate our participants' geographic life-space [13], which can be used to identify individuals in a community that may be at risk of not living a prosperous life or to determine early functional decline in older adults with a reduced life-space. Some of these results are not particularly astonishing from a medical point of view, but they were obtained through sensor data collected from their mobile phones. As opposed to conventional self-report methods, ambulatory assessment methods aim at measuring certain aspects while the participant undergoes normal daily activities. Therefore, mobile phones can be used for ambulatory assessment, which can support research in this area and provide reliable, ecologically-valid data.

### *13.2.2 Detecting Anxiety in Caregivers to Support Cognitive Behavioral Therapy*

Caring for people with dementia (PwD) is a demanding and stressful activity that frequently causes anxiety and might lead to depressive disorders in caregivers. Cognitive behavioral therapies can assist caregivers by providing coping strategies, such as breathing exercises or seeking social support. Some of these strategies can be more successful if they are enacted when the person is experiencing anxiety, but even the caregiver might not realize that she is experiencing anxiety or its consequences, which is one of the main drawbacks of using subjective ratings based on questionnaires and self-report. Sensors in smartwatches can be used to obtain physiological data, which can be used to infer state anxiety when the subject is experiencing a stressful situation. This could trigger coping strategies to reduce anxiety and improve the caregiver-PwD relationship.

We conducted a study, using the naturalistic enactment technique, in which 10 subjects were asked to care for an older adult who acted as if she was experiencing dementia [14]. Each subject cared for the PwD in 3 sessions, one per week, which lasted approximately 30 min. We used wearable devices to record the following physiological signals from the participants: Galvanic Skin Response (GSR), Heart Rate (HR), and Electroencephalography. To establish ground truth, we analyzed the videos of the sessions and asked participants to take notes during the session including their level of perceived anxiety. The physiological signal was processed in periods of 30 s to calculate 9 features from GSR and HR data. We obtained an average precision of 78% when recognizing two possible states: "Anxious" and "Not anxious", using a Support Vector Machine classifier [15]. A Markov chain model was evaluated using Inter-beat Interval data obtained from the HR signal, to detect 4 internal states: "Relaxed", "Arousing", "Anxiety", and "Relaxing". The average accuracy obtained was 73%.

The results provide evidence that the experiment elicits state anxiety and that it can be detected using wearable sensors. While the analysis was conducted in short intervals, triggering a coping strategy would normally be done after accumulated evidence has been obtained from a few minutes of data, reducing the probability of incurring in false positives, or missing false negatives.

### *13.2.3 Monitoring Eating Behaviors from Photographs Using Crowdsourcing*

One aspect of interest for aging research is monitoring food intake, given the implications that may have for several metabolic disorders and diseases diagnose and treatment. In particular, beyond identifying food that may be detrimental to the patient's health and wellness, monitoring nutritional content and caloric intake can be useful for clinical assessments. Advances in automatic behavior recognition and monitoring have fostered the development of computing systems aimed at supporting behavior change. Eating behaviors of interest might include eating late at night, prolonged fasting, or regular fast food consumption.

Monitoring eating behaviors by nutrition specialists has been traditionally addressed with various strategies including: a) paper-based logs and computational systems that facilitate meal logging, and b) automatic recognition approaches. Paper-based methods for monitoring eating behaviors include food records, food frequency questionnaires, or forms that include a meal description and time of intake. Mobile phones have been increasingly used to support this task, with apps to help individuals record their meals and coach them on healthful habits, such as Noom (noom.com) or Calorific (calorificapp.com). Regardless of the method used, both of these approaches rely on self-report, meaning that individuals must explicitly enter detailed data regarding their food intake on a daily basis. In the long run, this can be burdensome for individuals being monitored. Automatic recognition approaches are desirable due to low overhead but are of little practical use due to low accuracy.

To monitor food intake, it is desirable to burden subjects as little as possible without sacrificing accuracy. Burden-wise, an approach that is based on taking photos of the meals can be very convenient. Since high-precision automatic recognition is still underway, photos of meals can then be collectively analyzed by a crowd, who can help assess nutritional content or caloric intake. Still, knowing what type of approach to use for assessment to maximize accuracy and reduce burden in those rating photos is not a trivial question, for which we proposed and evaluated six assessment approaches (see Table 13.1), which were compared regarding Latency (time to make an assessment), Cognitive Load in raters, and Accuracy by raters.

From our results (see [16] for further details), in terms of Latency, participants required, on average, less than 30 s to assess one photograph across all approaches. Participants took less than 7 s, on average, to asses A4 and A3. In contrast, A5 was

**Table 13.1** Six approached proposed for assessing nutritional content or caloric intake

| Code | Approach | Description |
|---|---|---|
| A1 | Number of calories | The individual estimates the number of calories contained in a meal, from looking at the photograph |
| A2 | Food groups | The individual selects the food groups perceived from the photograph. The participant has to estimate the quantities (none, some, adequate, plenty) of each of the following food groups: fruits, vegetables, cereals, legumes, and animal origin. This is based on the Official Mexican Standard NOM-043-SSA2–2005 for a balanced diet |
| A3 | Healthfulness scale | The approach is designed to assess the healthfulness of food in the images. The user rates the photograph in a scale from 1 (not healthful) to 7 (very healthful) |
| A4 | Caloric range | This approach is similar to A1, calories in the meal shown in the photo are estimated by selecting one of six 200-cal intervals (e.g., 401–600) |
| A5 | List of ingredients | The user types in all the ingredients that she thinks are contained in the meal shown in the photograph, even those not in sight such as salt or cooking oil. |
| A6 | Similar images | The participant selects from a set of 9 different images the one she believes is the most similar to the photograph presented. This action is repeated twice, the second set of images depends on the first image selected. |

the one with highest latency (28.02 s). Regarding Cognitive Load, A1 scored the highest cognitive load. Conversely, A3 and A6 were perceived as the tasks with less cognitive load. Finally, regarding Accuracy, A1 and A4 had a low accuracy score, and A5 and A6 registered an acceptable accuracy, which improved with additional answers from the crowd.

Based on a modified version of A5, we developed Lucy a digital assistant aimed at helping patients undergoing weight-reduction treatment. We worked closely with a nutritionist clinic, in which doctors and nurses consult around 100 patients in terms of nutrition. We evaluated Lucy with patients of the clinic. Both Lucy and the crowd assessment have the potential to facilitate the work of nutrition experts in coaching numerous patients 24/7. Potential users were positive about adopting Lucy for future use.

## 13.3 Data Quality in Behavior and Activity Monitoring for Aging Research

Studies in mobile sensing for healthcare produce large, complex datasets with information opportunistically gathered from distributed sensors in mobile devices. This raises issues regarding the organization and sharing of the large amounts of data collected. Some of these issues include the heterogeneity of the devices, diversity of sensors used, and the need for data provenance when integrating datasets from diverse studies. Assessing quality is of paramount importance for conducting

**Table 13.2** Data management and collection issues for mobile sensing

| Category | Issue | Responsible |
|---|---|---|
| Research | Heterogeneity in data gathering | Researchers |
| | Data annotation | Researchers |
| | Data pre-processing | Researchers |
| | Limited data sharing | Researchers |
| Legal | Privacy | Participants/Researchers |
| | Ethics | Researchers |
| Engineering | Heterogeneity in sensor data | Hardware vendors |
| | Dispersion | Researchers/stakeholders |

longitudinal studies and building on historical knowledge as new data become available.

In mobile sensing for healthcare and aging research one of the main concerns is generating reliable datasets that can be used to push forward the boundaries of the area, which necessarily involves providing structure to the data. While current efforts have yielded promising results, we believe that scientists could benefit from a distributed repository of aging datasets. These datasets are to be curated and integrated to facilitate conducting new research in aging such as generalizing previous findings when comparing with new data from a different population and conducting longitudinal studies controlling for the conditions in which data were gathered over long periods of time.

In Table 13.2 we present some of the data management issues currently faced by the research community in mobile sensing for healthcare. These issues need to be addressed if a data infrastructure from which to continuously construct new knowledge and validate previous findings is to be made available. This is particularly important in aging research involving mobile or wearable devices, as the issues are to be increasingly present.

One issue that is of great concern for the development of aging research is that of preserving the privacy of participants, particularly since behavior and healthcare information are being recorded. Mobile and wearable devices collect large amounts of data from individuals and families. Although certain computational methods can be applied to data to remove sensitive information, it has been shown that with as little as the date of birth and zip code, one individual can be identified with a great degree of certainty.

One technical solution to partially address privacy concerns is for all data processing to be processed on the mobile device (e.g., mobile phone), but cost in terms of battery and processing power can be high. Other approaches involve data anonymization and encryption. Ultimately, addressing privacy involves considering social and technical approaches, and developing proper regulatory frameworks.

## 13.4 Conclusions

The emergence of these novel forms of ICTs offer unprecedented opportunities for scientists and practitioners to better understand and potentially influence patients and their contexts. These advances are being helpful in stepping up our understanding of human dynamics and contexts. Even more, a better understanding comes with an opportunity to provide adequate services to patients and potentially influence their attitudes and behaviors.

At the same time, these emerging technologies pose new challenges for well-established research methods as some of them will have to be rethought, reshaped, or overhauled. Technologies that could provide a continuous data stream of measurements for researchers and physicians were unthinkable two decades ago. Modern mobile and wearable devices, coupled with their augmented capabilities and ubiquity, are indeed creating new opportunities for measuring behavior, conducting epidemiological studies, and enacting interventions to change behavior. Mobile and wearable devices can boost medical studies by gathering data from larger populations, increasing the frequency of reporting and providing more reliable data based on the continuous monitoring of actual behavior rather than from sporadic interviews that rely on self-report. Analysis from larger populations, for instance, could result in findings that take into account differences in groups regarding gender, age, or upbringing, with results drawn rather rapidly and with higher ecological validity.

Finally, when comparing these emerging technologies to traditional approaches in research, there are stark differences between the types of data that can be collected (e.g., audio, video) as well as the frequency in which measurements can be taken (e.g., continuous), all of these with higher ecological validity. Nevertheless, there is plenty of work ahead to supplement current methods, rethink some of them, or perhaps create new methodologies that can take full advantage of the opportunities being offered to aging research by technology. Still, one of the big questions that will need to be addressed is whether revising current methods (and instruments) by increasing the frequency of measurements, increasing data quality, aggregating data from various sources, and having higher ecological validity will have, ultimately, medical significance. That is, if any improvements to current methods or new proposed methods will have an ultimate effect on the health outcomes of patients. Undoubtedly, incorporating these emerging technologies to research on aging will provide a different lens through which researchers, practitioners, and family members can scrutinize subtle changes in patient conditions and behaviors, perhaps before it is too late. Ultimately, what technology will surely provide, as it has been shown in other disciplines, is an increase of the scope, depth, and complexity in the design of much more comprehensive studies in the area.

## References


1. Ballinger B, Hsieh J, Singh A, Sohoni N, Wang J, Tison GH, et al (2018, Feb 2–7) PletcherDeepHeart: Semi-supervised sequence learning for cardiovascular risk prediction. AAAI conference on artificial intelligence (AAAI-18), New Orleans, Lousiana, USA2018
2. Larson EC, Lee T, Liu S, Rosenfeld M, Patel SN (2011) Accurate and privacy preserving cough sensing using a low-cost microphone. 13th international conference on ubiquitous computing (Ubicomp 2011): ACM, pp 375–84
3. Lu H, Frauendorfer D, Rabbi M, Mast MS, Chittaranjan GT, Campbell AT, et al (2012) StressSense: detecting stress in unconstrained acoustic environments using smartphones. Proceedings of the 2012 ACM conference on ubiquitous computing (Ubicomp 2012), Pittsburgh, Pennsylvania. 2370270: ACM, pp 351–360
4. Dong W, Lepri B, Pentland A (2011) Modeling the co-evolution of behaviors and social relationships using mobile phone data. 10th international conference on mobile and ubiquitous multimedia (MUM 2011); Dec 7–9; Beijing, China. 2107613: ACM, p. 134–43
5. Ferreira D, Kostakos V, Dey AK (2015) AWARE: mobile context instrumentation framework. Frontiers in ICT 2:6
6. Castro LA, Favela J, Quintana E, and Perez M (2015) Behavioral data gathering for assessing functional status and health in older adults using mobile phones. Personal Ubiquitous Computing 19(2):379–391. http://dx.doi.org/10.1007/s00779-014-0825-9
7. Beltran J, Navarro R, Chavez E, Favela J, Soto V, Ibarra C (2014) Detecting disruptive vocalizations for ambient assisted interventions for dementia. In: Pecchia L, Chen LL, Nugent C, Bravo J (eds) Ambient assisted living and daily activities. Springer International Publishing, Cham, pp 356–363
8. Fried LP, Tangen CM, Walston J, Newman AB, Hirsch C, Gottdiener J et al (2001) Frailty in older adults evidence for a phenotype. J Gerontol Ser A Biol Med Sci 56(3):M146–MM57
9. Rockwood K, Fox RA, Stolee P, Robertson D, Beattie BL (1994) Frailty in elderly people: an evolving concept. CMAJ 150(4):489
10. Clegg A, Young J, Iliffe S, Rikkert MO, Rockwood K (2013) Frailty in elderly people. Lancet 381(9868):752–762
11. Katz S, Ford AB, Moskowitz RW, Jackson BA, Jaffe MW (1963) Studies of illness in the aged: the index of ADL: a standardized measure of biological and psychosocial function. JAMA 185(12):914–919
12. Ware JE, Sherbourne CD (1992) The MOS 36-item short-form health survey (SF-36): conceptual framework and item selection. Med Care 30:473–483
13. Schenk AK, Witbrodt BC, Hoarty CA, Carlson RH, Goulding EH, Potter JF et al (2011) Cellular telephones measure activity and lifespace in community-dwelling adults: proof of principle. J Am Ger Soc 59(2):345–352
14. Miranda D, Favela J, Ibarra C, Cruz N (2016) Naturalistic enactment to elicit and recognize caregiver state anxiety. J Med Syst 40(9):192. https://doi.org/10.1007/s10916-016-0551-0
15. Miranda D, Favela J, Arnrich B (2017) Detecting anxiety states when caring for people with dementia. Methods Inf Med 56(01):55–62. https://doi.org/10.3414/ME15-02-0012
16. Parra MO, Favela J, Castro LA, Morales A (2018) Monitoring eating behaviors for a nutritionist e-assistant using crowdsourcing. IEEE Comput 51(3):43–51. https://doi.org/10.1109/MC.2018.1731078

# Chapter 14
# The Challenge of Big Data and Data Mining in Aging Research

**Juan Carlos Gómez-Verján and Luis Miguel Gutiérrez-Robledo**

*"Indeed, we are at the point in history where Big Data should no intimidate, but inspire us".*

*-Francis Collins-*

**Abstract** Population growth and fast evolution of several technologies into almost every corner of human activities have become a constant in the modern society nowadays. Consequently, an enormous amount of information is generated every day, from many sources, such as: web information, mobile phones, social media, scientific reports, medical and healthcare information among others, resulting in the so called Big Data. Digitization, storage, collection, and particularly analysis of patterns on these data have led to advances in data sciences (Data Mining) to obtain valuable information from people necessities, demands, hobbies, and activities daily from such databases (Big Knowledge). On the other side, aging has become one of the main challenges of modern societies, since there is an increase on prevalence of chronic pathological conditions and aging associated outcomes. Biomedical and clinical research have experienced several advances in the last years with the implementation of omics technologies and digitization of healthcare, in this context, Big Data and Data Mining become a powerful tool to obtain valuable information that could be used to approach the complexity of the aging process and the clinical implications for older adults. Several companies and governments all over the world have already started to successfully implement these types of technologies from pharmaceutical developments to public health issues. Although future seems promising novel holistic approaches are needed to implement a multidisciplinary agenda

J. C. Gómez-Verján (✉)
Department of Basic Research, National Institute of Geriatrics, Mexico City, Mexico
e-mail: jverjan@inger.gob.mx

L. M. Gutiérrez-Robledo (✉)
National Institute of Geriatrics, Mexico City, Mexico
e-mail: lmgutierrez@inger.gob.mx

 
C. García-Peña et al. (eds.), *Aging Research - Methodological Issues*,
https://doi.org/10.1007/978-3-319-95387-8_14

that could help us understand the complexities of human aging and develop the appropriate interventions to improve older people's health in healthcare systems.

**Keywords** Big data · Data mining · Big knowledge · Systems biology · Public health

## 14.1 Big Data

Along with the population growth worldwide, and the fast evolution of technologies into almost every corner of human activities, an enormous quantity of information is generated every day, from many sources, such as: web information, mobile phone usage, social media activities, consumer preferences, financial systems, climate information, scientific reports, medical and health care information [1], moreover, the software and data business company Domo Inc. on its annual report of the world's data generation called "Data Never Sleeps 5.0" for 2017, indicate that there are almost 3.7 billion users in the global internet population, and there is an output of 2.5 quintillion bytes a day of data created daily [2]. In this context, the digitalization, storage, collection, maintenance and analysis of these data have led to advances in data sciences and into the so called "Big Data Era". This revolution has led to changes in prevalent paradigms, since most of these data cannot be managed by traditional and conventional techniques for data management, and novel infrastructure and approaches for data sciences need to be taken by executives, academics and economist all over the world.

John Mashey a researcher form Pennsylvania University in computer science was responsible for making the term "Big Data" quite popular in the early 1990's decade. Nowadays, Big Data is usually defined as huge amount of data (in the order of Terabytes-1012 bytes and Petabytes-1015 bytes) highly complex, that within a reasonable period of time cannot be captured, managed, processed, interpreted and organized as information that could be read by human beings or traditional data processing software [1]. Most of the technologies used for these novel approaches are computational linguistics and machine learning. Table 14.1 indicates some of the main software available for the Big Data processing and management. In this context, accordingly to Davenport, Big Data could be classified in two main groups: machine-generated (data created by machines without human intervention) and human-generated (created with human intervention).

### 14.1.1 Data Mining and Big Data Analysis

Managing data is one of the most important issues once you start to work with Big Data, several authors have stated different methodologies to start to work with, that in general could be divided by: Data collection, Data measures, Data Analysis and Knowledge discovery [3], accordingly to: Volume (data generation and processing

**Table 14.1** Main software available for big data and data mining processing

| Software | Characteristics | Type |
|---|---|---|
| *Apache Hadoop* | *Is an open-source software for the processing, and storage of big data, using MapReduce programming models on computer clusters, therefore, there is no problem with hardware, it possess the ability to handle limitless concurrent jobs or task* | Open source |
| *HPCC systems big data* | *This is a platform for transforming, querying and data warehousing, implemented on commodity computing clusters to provide high-performance on data parallel processing for big data* | Open source |
| *Mongo DB* | *Is a no-SQL database, written in C++, that helps to store and analyze big data* | Open source |
| *R-programming* | *R programming language, has been designed and implemented as a data mining tool and is widely used for data analysis and machine learning applications, currently there are a series of packages available for big data* | Open source |
| *RapidMiner* | *Is a software platform that provides integrated environment for data preparation, machine learning and predictive analysis* | Commercial |
| *Knime* | *Is a leader in the analysis (machine learning), integration and reports for big data, it possess an open source version available for public in general and a quite remarkable commercial version for specialized data scientist* | Open source and commercial |
| *Orange* | *Is an open-source data visualization, machine learning and data mining highly used tool that possess algorithms for predictive modelling it could be used as a Python library* | Open source |
| *Spark* | *Is an open-source cluster-computing framework that provides an interface programming for big data analysis, maintenance and analysis.* | Open source |
| *IBM SPSS modeler* | *Is a data mining workbench for analyzing and developing predictive models suppported by IBM SPSS software* | Commercial |

numbers), Velocity (data processing in concordance with data generation speediness) and Variety (quality of data i.e. type of data and if it is structured or unstructured).

Data Analysis and Knowledge discovery will make us get novel and useful information from the databases that could be useful for prediction, classification and innovation depending on the type of information we are working with. In this context, the so called Data Mining defined as an analytic process designed to discover or extract patterns, novel information and systematic relationships from variables of large data sets (usually Big Data) [4], involving also the process of storage and processing the data, as well as the complicated process of presenting the results in a way understandable and easily interpretable for everybody The last goal of Data Mining is the prediction of patterns, the term was first coined by Piatetsky-Shapiro and Frawley in the early 90's when there was a rush for developing novel algorithms for data processing in business by several software and data companies all over the world [5], it was originally called Data Collection by IBM in the 60's. Data mining techniques are the result of a long process of research in product and business development, that began when the business of data warehousing and collection began to grow.

In order to make Data mining usable as an important tool it involves the complete understanding of the architecture of the data and the analytical methods to be used, since predictive relationships of data may not necessarily represent causes of an action or a behavior [6]. Although a detailed review of the complete Data mining technique is beyond the scope of this chapter, briefly, the process of data mining consists in 3 different stages [7]:

A) Exploration and Collection of data: At this stage of the analysis we usually focus on the type of data we are going to work with, the type of data, and the nature of the analytical problem we wish to work with. Usually this stage involves the process of data curation [8], i.e. selection of adequate data with complete meta-information for the analysis, transforming data to the correct format for the software analysis, type of variables and predictors to be measured and usually exploratory analysis.
B) Modelling the data: Building a model involves the knowledge of the kind of answers you wish to apply or the type of prediction you wish to do; since there are several software packages with already known models developed [6]. This stage is critical and involves the highest and most elaborated process in Data mining, in this context, there are a wide variety of techniques, such as: association, normal statistics, classification, Bayesian statistic, and several machine learning techniques [9] such as: clustering, neighborhood, decision trees, neural networks, deep learning, bootstrap aggregating, boosting, etc.
C) Deployment and validation of the data: Building the model is not the end of the analysis, and once the model has been processed and evaluated, you must validate and interpret the results, their significance, and in case how does the novel information could be classified, remembering that Data mining is not statistics, we do not care how the data is distributed but what is the potential of such data for prediction and, or pattern identification [10]. Moreover, if you gained knowledge from the data you will then need to organize and present such results in a way that anyone in the field could use it. In this stage, there are several examples of informatics tools useful for deployment of the data and the exporting of algorithms, such as: predictive model markup language, portable format for analytics, confusion matrices, SQL, R-algorithms, Python, applications for Java and several programming tools [11].

## 14.2 Big Data and Health

It is widely known that health involves the production of an enormous quantity of data including diagnostic results and images, medical records, laboratory results, public health registry in the different levels, and data produced for biomedical and for clinical research [12]. In this context, Ruckenstein et al. [13] proposed the so-called "datafication of health" i.e. the conversion of all aspects of health at its different levels, into quantifiable data, meaning that all techniques that are currently being used for Big Data and Data mining could be applied to analyze health at different levels, the so called "Biomedical Big Data" (BBD). Moreover, the value of health research based on non-traditional data streams services from internet such as: e-mail, online purchasing and video conferences has already been demonstrated [14].

BBD could help us analyze and store people's information throughout their lives on: diseases, phenotype, genotype, behavior, environmental location, occupation, and clinical data, making therefore, health-predictions easily for individuals and in consequence for populations [12]. In this context, it could provide of the correct tools to governments, public health departments, and decision makers for the implementation of the adequate prevention politics and interventions, in order to improve health on population, for instance, machine learning techniques have allowed several medical disciplines to create more accurate prognostic and diagnostic models based on pattern recognition (computer-aided detection), that could help physicians to improve their diagnosis [15]. Several governments all over the world have started to invest and create departments dedicated only to the analysis of BBD, for instance, the National Institutes of Health (NIH) launched the "Big Data to Knowledge" (BD2K) initiative in 2012 [16], which involves multiple research centers such as the Big Data for Discovery Science Center and the Center for Expanded Data Annotation and Retrieval, as well as a set of focused individual research and training projects, to enable biomedical research and consider novel approaches in data science, to facilitate discovery and support new knowledge, the most important objective of this initiative is to index software to operate these datasets; on the other hand, the e-Health Action plan for 2012–2020 for the European Commission of Health, has allocated 2 billion € under Horizon 2020 program to invest in research and innovation on Big Data [17] as well as, a public-private partnership with a budget of 1.638 billion from the EU commission and 1.425 from other life sciences industries and organizations. Moreover, an Oxford Economics revealed that at least 70% of healthcare companies are looking to invest in Big data, Data mining and cloud computing expecting a significant impact in innovation in several topics on healthcare field. In this context, as stated by Vayena et al. [14], it is important to think of health-related big data as an evolving ecosystem.

## 14.3 Big Data in Epidemiology

Epidemiology and public health, have changed dramatically over the last years, and are now more interconnected with so many other sciences due to the global advances in technology and to the interdisciplinary nature of such sciences. Nowadays, the use of Big Data for epidemiologist is as referred by Salerno J., et al., "the exploration and interpretation of very large and complex datasets derived from pooling cohorts, from omics projects, electronic stored medical records, and health digital information" [18], since usually a normal epidemiologist turns for primary and secondary data sources to initiate its study, however, in the Big Data era, there is an enormous amount of information and sources of data (medical records, biobanks, geolocalization, shopping habits, genomic data, pharmaceutical prescriptions, social behavior, among others). In this context, novel epidemiologist must be prepared for time-consuming data collection, curation and storage, as well as, novel approaches using Data Mining tools and statistical approaches, as well as novel ethical and legal challenges related to potential harms of Big Data use, including confidentiality and privacy issues and other potential concerns from institutions and public agencies all over the world.

Several examples of epidemiological studies and Big Data have arisen over the last years, among the most important projects is the Nordic Arthroplasty Register Association (NARA) database which includes information concerning implant brands, fixation methods, and implant survival,, since 1995 in Denmark, Finland, Iceland, Norway, and Sweden, generating information from more than 1 million patients from each of the different countries [19] in each of the countries; other interesting example is the Observational Medical Outcomes Partnership (OMOP) which is a public-private partnership formed by several public and private representatives including the Food and Drug Administration (FDA) and members of the Big Pharma industry mainly focused on pharmacoepidemiology using and incentivizing novel healthcare databases through electronic health records allowing therefore, linkage possibilities and the possibility of a lifelong complete follow-up [20]; moreover, South Korea which is a world leader in information technology infrastructure launched in 2011 the so-called Big Data Initiative, which established a pan-governmental big data network and analysis systems, it was so important that in 2014 the ministry of science of such country released the Medical information consulting program to collect medical data and customize treatment and help the national health insurance service become more efficient, providing information to patients such as: duration of illness, cost of treatments, medical services, cases per location, institutions specialized in diseases; and to medical industry of pharmaceutical trends, distribution of drugs, medical equipment and devices most asked for by the population [21]. A very successful case example was the Seoul National University Bundang Hospital, which was the first hospital in the Asia-Pacific region to fully digitalized big data, and doctors and nurses are able to configure systems with precise clinical information, improving time of patient referral from 48 h to 4–6 h, and for example reduce the dosage of antibiotics before surgery [22].

## 14.4 Big Data Biomedicine Research and Omics

It has been years since the Human Genome Project results were published, and biology experienced a revolution; as well in our understanding of biological mechanisms as well as in access to an enormous amount of biomedical data now freely

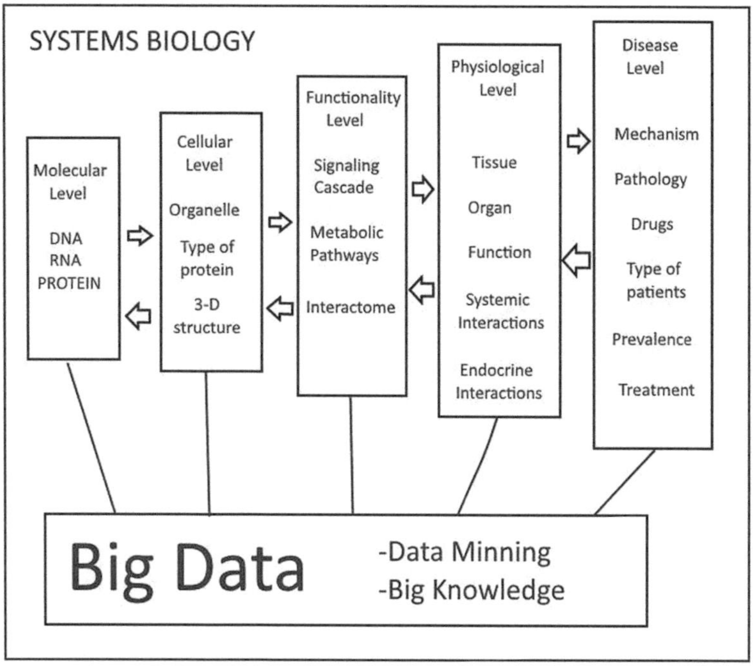


**Fig. 14.1** Holistic points of view in systems biology. Analysis coming from different omics technology could only be helpful if we have a holistic approach from different levels this is the objective of systems biology

available. Novel -omics tools are being daily applied to several topics in biomedical research projects in the last few years (genomics, transcriptomics, proteomics, epigenomics, metagenomics, metabolomics, and microbiome, among others) [23]. Moreover, this information now helps us understand relationships between genotype, phenotype, behavior, and many outcomes of human beings through the development of the so-called Systems Biology (Fig. 14.1) [16].

However, information obtained from such technologies increases dramatically, as well as the need of developing public repositories. Besides, each Omic experiment represents challenges, since standard nomenclature and analysis for genes are not the same for proteins or metabolites. Each technology involves the production of different data in different formats (Table 14.2), standards for information and meta-information are needed so that stored data is clear enough for interpretation and re-use from other scientist to compare and reproduce experiments [24].

In the last few years, there has been an international effort from several consortia all-over the world to develop and standardize -omics technologies results to be used and published e.g. for Genomics (Minimum Information about a Genome/Metagenome Sequence from Genomic Standards Consortium), Transcriptomics (Minimum Information about a Microarray Experiment from Functional Genomics Data Society), Proteomics (Minimum Information about a Proteomics experiment

**Table 14.2** Main omics used in biomedical research and most common formats

| Omics | Technologies used | Formats |
|---|---|---|
| Genomics | Genotype microarrays, DNA-seq, Whole genome sequencing, Whole exome sequencing | FASTA, FASTQ, SAM, BAM, VCF, CEL, CSV, GFF, BED, LOH, SNP, IGV, MUT, TSV, JSON |
| Transcriptomics | RNA-microarrays, RNA-seq, SAGE-seq | FASTA, FASTQ, BAM, SAM, CEL, RES, CSV, GFF, GCT, GFF, BED, IGV, SEG, TSV, JSON |
| Proteomics | Protein chips, Mass spectrometry, 2-D gel electrophoresis, Mononuclear-NMR | mzML, TraML, mzIdentML, mzXML, mzData, mzQuantML, MSF, tandem, omx, dat, FASTA, PEFF, mzTab, PRIDE XML, MGF, ms2, pkl, TSV, CSV |
| Epigenomics | ChIP-Seq, MeDIP-Seq, DNase-Seq, Mnase-Seq | FASTA, FASTQ, SAM, BAM, CSV, TSV, GFF, SEG, BED, WIG, txt |
| Metabolomics | Mono and heteronuclear-NMR, Mass spectrometry, Chromatography and mass spectrometry, | CDF, mzML, mzXML, mzData, mzQuantML, txt, TSV, CSV, CVR, VSN, mzIdentML |

from the Human Proteome Organization), Metabolomics (Core information for Metabolomics), Epigenomics/Transcriptomics (Minimum information about a high-throughput Nucleotide sequencing Experiment by Functional Genomics Data Society) [24]. In this burgeoning context, information will soon be reproducible and amenable to knowledge translation into personalized medicine.

## 14.5 Challenges of Big Data and Aging

It is widely known that life expectancy all over the world has increased in the last few years. According to the World Health Organization report on Aging and Health, in 2016 global life expectancy for females is 73.8 and 69.1 for males [25] as life span grows, the prevalence of chronic conditions and age associated disease increases, and becomes a matter of interest for several public and private stakeholders. Multi-morbidity results from this phenomenon and represents a challenge for health care systems. Diversity increases as we age, and disadvantage plays a significant role in modelling these differences, which vary widely throughout a lifetime. In this very complex and diverse context, recent advances on information, communication and biomedical technologies such as electronic devices, medical digital diagnostics and prescriptions, medical assistive and wearable devices, personalized medicine and genomics, among others, represent an opportunity for an efficient use in order to provide the health-care system with new tools based on anticipation, prevention and an improved continuum of care. Besides, another field of development, as electronic medical records improve and become widespread; could be the use of a Big Data approach to provide decision makers in the public and private sectors with novel information for planning interventions and public health policies;

and to help develop new pharmaceutical and health device resources answering to the needs of an aging population [26, 27].

Modern developments in Big data and Data Mining represent an opportunity to improve multidisciplinary exchanges "Big Knowledge" and accordingly to William Callaghan a novel paradigm in the scientific method, since id. "...comprehensive data coverage unearths causal relationship between phenomena…" [28], which is not far away from most of modern scientific projects for instance, most genomic projects involve first the genome acquisition and consequently the data analysis to develop a hypothesis. However, since Big Data involves a holistic approach multidisciplinary research groups must be created so that results could improve in order to raise the probability of moving towards the right conclusion especially in the field of aging, where a continuous monitoring and correct interventions, as well as prevention are among the main challenges for health systems [26]. However, we must be aware of the fact that already, among the complexity of aging in modern life; the lived reality of big data should be approached with caution because all the data we are shedding every day is too revealing of our intimate selves but may also misrepresent us. Like a fluorescent light in a dark corridor, it can both show too much and not enough [29]. Therefore, we do not really know the true potential or the dangers of big data analysis, and we must be restrained in our approach and look for a balance between enthusiasm and the reality of our limitations. It's important to mention that any government that adopts any type of Big-Data approach on public health must create policies for the protection of individual data and at the same time promote access and sharing for the use of such data for public benefit [14].

Over the past few years many long-term care programs have been implemented for aging in several countries such as: Japan, Taiwan, USA, Australia, Canada, South Korea, and Denmark, which intended to implement information and communication technologies to improve long-term care systems the so-called aging in place [26]. Even so, Japan's government made mandatory the long-term care insurance system and implemented the "Guideline to promote the Appropriate Use of Information Systems in Care for the Elderly in Conjunction with At-home Care" to stimulate Big Data networks of information from healthcare providers and licensed in home caregivers, nursing and doctors, as well as information from patients such as residence, medical treatment, nursing, etc. [30]. Another successful example of the implementation of Big Data to healthcare systems in older people, is the Australian e-Health Research Centre which have almost 50 researchers, software engineers and doctorate students developing technologies in partnership with clinicians to improve health care systems. They have developed an innovative platform for cardiac rehabilitation, using wearable devices and mobile phones [31]. Moreover, several start-up companies have moved into the field, such as the New York based Hometeam Inc. which developed its own software to match caregivers with families through mobile technology in a very personalized care planning for older adults. Another successful example is CareZapp Inc. located in United Kingdom; they developed a mobile app that creates an ecosystem platform connecting caregivers, volunteers, family, friend and doctors to improve the health of older patients, by means of in-home sensors, wearable devices, and mobile phone with notifications 24/7 of patient status [32].

## 14.6 Globolomics of Aging Research

For decades, researchers in the biology of aging have focused on defining mechanisms that modulate aging by primarily studying a single metric, sometimes described as the "gold standard" lifespan. Increasingly, geroscience research is turning towards defining functional domains of aging such as the cardiovascular system, skeletal integrity, and metabolic health as being a more direct route to understand why tissues decline in function with age (see Chap. 4 on Geroscience). Each model used in aging research has strengths and weaknesses, yet we know surprisingly little about how critical tissues decline in health with increasing age and how the different systems interact. We know very little as well about the interplay between the biological mechanisms of aging and chronic disease.

Over the last few years, research on the molecular foundations of aging and consequently anti -aging therapies, have increased dramatically, just to mention an example, a search for the words: "molecular" and "aging" in PubMed database gets almost 35,563 documents, increasing from 553 documents in 1997 to 3551 in 2017. It becomes clear that several Omic approaches have been critical for progress in the field, performed either with clinical human samples or with animal models (*C. elegans, S. cerevisiae, M. musculus, Rattus norvegicus* etc). In this context, there is an enormous amount of information freely available in the web, consequently, Big Data and Data Mining approaches become relevant in order to systematically extract information and obtain knowledge from such.

As previously mentioned nowadays there are several Omics technologies available, however integration of the data is the main challenge in the field and remains a black box in genomic studies [33], however a new way to improve our understanding of the aging process is acknowledging its complexity and developing new methods capable of embracing it. The use of such technologies and its results, each omic tool, will only give an answer to specific different questions i.e. proteomics will give us information that may be or may be not be related with metabolomics. Therefore globolomics (or "deep phenotyping" leading to molecular or genetic epidemiology, albeit at a finer resolution) approaches must be consider when performing studies in aging so we could have a systems approach of results. Moreover, association with physiological conditions must be consider [34], projects such as the Human Physiome Project become increasingly important, as their analysis could give us a better approach to the proportion of genetic and environmental (nature and nurture) participation in aging and consequently, contribute to the development of a novel aging theory.

## 14.7 Expectations on Big Data in Aging

It's been quite a few years since the declaration of the Madrid International Plan of Action on Ageing, called for the elimination of the inequalities in health-access and to the development of novel health-care systems policies for older people [35], and although it's still a long and arduous road to travel, novel technologies on informatics, genomics and communication could help us diminish such inequalities and move forth in all the fields of aging research. In this context, there are a lot of international efforts already being successfully developed from private and public sectors, and a number of researchers and governments are already compiling aging and health information in current databases [30, 31, 36]. The more we learn on Big Data and Data Mining the faster we will close this gap.

## References

1. Yang C-T, Liu J-C, Chen S-T, Lu H-W (2017) Implementation of a big data accessing and processing platform for medical records in cloud. J Med Syst 41:149. https://doi.org/10.1007/s10916-017-0777-5
2. Data Never Sleeps 5.0 | Domo [Internet]. [cited 29 Nov 2017]. Available: https://www.domo.com/learn/data-never-sleeps-5
3. Vitari C, Raguseo E (2016) Digital data, dynamic capability and financial performance: an empirical investigation in the era of big data. Systèmes d'Information & Management 21(3):6392. https://doi.org/10.3917/sim.163.0063
4. Han J, Kamber M, Pei J (2012) Data mining trends and research frontiers. Data Min:585–631
5. Piatetsky-Shapiro G (1994) An overview of knowledge discovery in databases: recent progress and challenges. Workshops in computing pp 1–10
6. Maimon O, Rokach L (2009) Introduction to knowledge discovery and data mining. In: Data mining and knowledge discovery handbook, pp 1–15
7. McCue C (2015) Chapter 3 - data mining and predictive analytics. In: McCue C (ed) Data mining and predictive analysis, 2nd edn. Butterworth-Heinemann, Boston, pp 31–48
8. Freitas A, Curry E (2016) Big data curation. In: New horizons for a data-driven economy, pp 87–118
9. Kononenko I, Kukar M (2007) Chapter 3 - machine learning basics. In: KIK M (ed) Machine learning and data mining. Woodhead Publishing, pp 59–105
10. Michael A, AGSL B (2008) Mastering data mining: the art and science of customer relationship management. Ind Manag Data Syst 100(5):245–246
11. McCue C (2015) Process models for data mining and predictive analysis. In: Data mining and predictive analysis, pp 51–74
12. Vayena E, Blasimme A (2017) Biomedical big data: new models of control over access, use and governance. J Bioeth Inq 14(4):501–513. https://doi.org/10.1007/s11673-017-9809-6
13. Ruckenstein M, Schüll ND (2017) The datafication of health. Annu Rev Anthropol 46:261–278
14. Vayena E, Dzenowagis J, Brownstein JS, Sheikh A (2017) Policy implications of big data in the health sector. Bull World Health Organ 96(1):66–68. https://doi.org/10.2471/BLT.17.197426
15. Cabitza F, Rasoini R, Gensini GF (2017) Unintended consequences of machine learning in medicine. JAMA 318(6):517–518. https://doi.org/10.1001/jama.2017.7797
16. Bui AAT, Van Horn JD (2017) NIH BD2K centers consortium. Envisioning the future of "big data" biomedicine. J Biomed Inform 69:115–117. https://doi.org/10.1016/j.jbi.2017.03.017

17. Kolitsi Z, Thonnet M (2014) New directions in eHealth governance in Europe. In: Managing eHealth. Palgrave Macmillan, London, pp 50–60
18. Salerno J, Knoppers BM, Lee LM, Hlaing WM, Goodman KW (2017) Ethics, big data and computing in epidemiology and public health. Ann Epidemiol 27(5):297–301. https://doi.org/10.1016/j.annepidem.2017.05.002
19. Johanson P-E, Fenstad AM, Furnes O, Garellick G, Havelin LI, Overgaard S et al (2010) Inferior outcome after hip resurfacing arthroplasty than after conventional arthroplasty. Evidence from the nordic arthroplasty register association (NARA) database, 1995 to 2007. Acta Orthop 81(5):535–541. https://doi.org/10.3109/17453674.2010.525193
20. Stang PE, Ryan PB, Racoosin JA, Marc Overhage J, Hartzema AG, Reich C et al (2010) Advancing the science for active surveillance: rationale and design for the observational medical outcomes partnership. Ann Intern Med 153(9):600–606. https://doi.org/10.7326/0003-4819-153-9-201011020-00010
21. Jung JJ, Kim P (2017) Big data technologies and applications: 7th international conference, BDTA 2016, Seoul, South Korea, 17–18 Nov, 2016, Proceedings. Springer
22. Yoo S, Hwang H, Jheon S (2016) Hospital information systems: experience at the fully digitized Seoul National University Bundang hospital. J Thorac Dis 8(Suppl 8):S637–S641. https://doi.org/10.21037/jtd.2016.08.44
23. He KY, Ge D, He MM (2017) Big data analytics for genomic medicine. Int J Mol Sci 18(2):1–18. https://doi.org/10.3390/ijms18020412
24. Chervitz SA, Deutsch EW, Field D, Parkinson H, Quackenbush J, Rocca-Serra P et al (2011) Data standards for omics data: the basis of data sharing and reuse. Methods Mol Biol 719:31–69. https://doi.org/10.1007/978-1-61779-027-0_2
25. OECD (2017) OECD average life expectancy and perceived health, since 2005 [internet]. https://doi.org/10.1787/how_life-2017-graph10-en
26. Song P, Chen Y (2015) Public policy response, aging in place, and big data platforms: creating an effective collaborative system to cope with aging of the population. Biosci Trends 9(1):1–6. https://doi.org/10.5582/bst.2015.01025
27. Kwon Y, Natori Y, Tanokura M (2017) New approach to generating insights for aging research based on literature mining and knowledge integration. PLoS One 12(8):e0183534. https://doi.org/10.1371/journal.pone.0183534
28. Callaghan CW (2017) Developing the transdisciplinary aging research agenda: new developments in big data. Curr Aging Sci 10. https://doi.org/10.2174/1874609810666170719100122
29. Crawford K, Finn M (2014) The limits of crisis data: analytical and ethical challenges of using social and mobile data to understand disasters. GeoJournal 80:491–502. https://doi.org/10.1007/s10708-014-9597-z
30. Tamiya N, Noguchi H, Nishi A, Reich MR, Ikegami N, Hashimoto H et al (2011) Population ageing and wellbeing: lessons from Japan's long-term care insurance policy. Lancet 378(9797):1183–1192. https://doi.org/10.1016/S0140-6736(11)61176-8
31. Hansen DP, Gurney P, Morgan G, Barraclough B (2011) The Australian e-Health research centre: enabling the health care information and communication technology revolution. Med J Aust 194(4):S5–S7
32. Sonnega A, Robinson K, Levy H (2016) Home and community-based service and other senior service use: prevalence and characteristics in a national sample. Home Health Care Serv Q 36(1):16–28. https://doi.org/10.1080/01621424.2016.1268552
33. Lorusso JS, Sviderskiy OA, Labunskyy VM (2017) Emerging omics approaches in aging research. Antioxid Redox Signal. https://doi.org/10.1089/ars.2017.7163
34. Lund E, Dumeaux V (2008) Systems epidemiology in cancer. Cancer Epidemiol Biomark Prev 17(11):2954–2957. https://doi.org/10.1158/1055-9965.EPI-08-0519
35. Sidorenko AV, Mikhailova ON (2013) Implementation of the Madrid international plan of action on ageing in the CIS countries: the first 10 years. Adv Gerontol 26(4):585–593
36. de Magalhães JP, Stevens M, Thornton D (2017) The business of anti-aging science. Trends Biotechnol 35(11):1062–1073. https://doi.org/10.1016/j.tibtech.2017.07.004

# Chapter 15
# Research in Public Policies for Aging

Elizabeth Caro-López and Ernesto Velasco-Sánchez

**Abstract** Evidence-based decision-making is an imperative for conducting effective policies, particularly in the case of aging. Systematic research can contribute to better decision making by clarifying concepts, allowing us to place an issue into a larger theoretical framework and by providing evidence on what works. It can help to assess the efficacy, efficiency and legitimacy of different policy alternatives. Finally, it can inform of the potential problems that could be faced and present evidence of the effectiveness of the adopted policies. This chapter offers a brief review of the contributions that researchers can make at different stages in the policy cycle.

**Keywords** Public policy · Evidence-based policies

## 15.1 Introduction

A public policy is not a single action of government in response to specific political circumstances or social demands of the moment. Public policy has been defined in several ways; for the purposes of this chapter we will use one of the most comprehensive definitions. Public policy is everything the government does, and the main objective is to match goals to facts, purposes and mandates that it receives from society through a process that brings together resources, laws, organizations/institutions and programs [1].

The process of satisfying the needs that are thought to be in public interest is complex. Each need must be clearly identified and recognized and, ideally, it should have a minimum level of social consensus to merit public action. There are aspects of social development in which there is general agreement in favor of public action, such as the provision of public services, education, health, and public safety, among

E. Caro-López (✉)
National Institute of Geriatrics, Mexico City, Mexico
e-mail: ecaromx@gmail.com

E. Velasco-Sánchez
CIVICUS Consultants, Coyoacán/Mexico City, Mexico
e-mail: ernestovelascos@gmail.com

 
C. García-Peña et al. (eds.), *Aging Research - Methodological Issues*,
https://doi.org/10.1007/978-3-319-95387-8_15

others. New areas of development have been included more recently in governmental action, such as gender equality, promotion of a sustainable environment and, without a doubt, the aging population.

In today's societies there are increasing interests and needs to be attended to, and the most recent discussions include not only the list of needs to be addressed but mainly the priorities, as well as the forcefulness and immediacy in which they can and must be attended to.

What priority does attention to age and aging have in the actions of institutions? What economic, material and human resources are assigned to them and how should they be assigned? What are the costs and the benefits? Are the interventions defined by the evidence? How are these interventions evaluated?

All these questions reveal that public policy on a complex issue like the aging of the population requires two interrelated approaches: one political-normative approach that consists of understanding and attending to needs that are concordant with the legal framework, and one techno-scientific that implies collecting enough evidence to enable the government and its institutions to set priorities and act efficiently and effectively.

The development and implementation of evidence-based public policies is a challenge worldwide, since most government decisions are not based on enough evidence or scientific-technical knowledge does not become available fast enough for the policy designers to be efficient. Another factor in the policy cycle can also explain the gap between the politically defined needs to be addressed and the technologies and actions implemented to do so.

This chapter aims to show the importance of a close relationship between the policy sciences and the decision-makers in the field of age and aging. In particular, how can research contribute to better policies in this area and relevant contributions to different stages of the policy-making process? In Sect. 15.2, an overview of the policy-making process is offered, with attention to the usefulness of research in each stage. Section 15.3 discusses the challenges that have to be overcome in order to increase the use of research in decision-making.

## 15.2 The Process of Public Policies

Conventionally, policy-making has been looked at as a process, meaning a series of interconnected decisions that aims to solve or manage a public issue or problem. The most popular model is the policy cycle that depicts policy-making as a series of tasks, each one producing elements necessary to perform the next, resulting in a feedback loop [2]. There are several versions of this model, and in the following paragraphs, we explain each stage briefly.

### *15.2.1 Establishing Public and Government Agendas*

An agenda is a set of issues or challenges that are considered to be important. The public agenda is shaped by the convergence or social agreement on the salience of an issue. The government agenda is made up of the issues included in the social agenda that public officials decide to act upon. This decision is influenced by the priorities of the political elite and the representatives of the people (i.e. members of Congress).

At present, the issues of age and aging of the population are on most public and government agendas both nationally and internationally. However, the understanding and level of attention given to the issues involved can vary widely among different stakeholders and the manifestation can also differ from country to country and even from region to region within a single country. For the international community, represented by the United Nations (UN) and the World Health Organization (WHO), the issue is mainly of interest because of its implications on health, social development and human rights. In recent years this interest has increased, not only due to the growth of the population 60 years of age or older, but also because of the conditions in which people are aging and the implications for family, society and governments.

The UN has shown interest in the issue of older adults since 1948 when the General Assembly approved Resolution 213 (III) relating to the project for the Declaration of Old Age Rights. Since then it has been approached indirectly by the General Assembly and by organisms interested in social issues [3]. It was not until 1977 when the problem was put forth directly with emphasis on the need to organize a world assembly on older adults, and in 1978 it was agreed that this conference would take place in 1982. The Second World Assembly on Ageing took place in 2002. These meetings have resulted in a large number of documents that propose different approaches to dealing with the issue. WHO has always been concerned with this issue, but after 2000 it began to promote specific studies such as the Study on Global Aging and Adult Health (SAGE). Another example of how the question has become more relevant is the fact that April 7th of 2012, World Health Day, was dedicated to the topic of aging and health.

There seems to be a growing consensus as to the importance of the issues of aging and adult health, seen from a wider perspective that involves the health sciences, but also aspects such as human rights, human development and equality.

### *15.2.2 Defining the Issue*

If there is one crucial aspect in defining a public policy it is the definition of the problem itself. Techno-scientific evidence is of vital importance at this stage. In the case of aging, there seem to be sufficient technical and scientific reasons to argue the important of paying attention to this issue, as there seems to be a worldwide and local consensus that the population is getting older and that this has implications in society. However, defining a public policy problem is not as simple as it may seem: an agreement is needed as to what the core characteristics are, age and aging itself, and what preventive or care alternatives are available in order to avoid situations such as frailty, sarcopenia, disability, dependency or the geriatric syndromes. What are the dimensions and causes? Who is affected and to what extent? And, how will this situation pan out if nothing is done about it? Does the government have the capacity to deal with these issues? There are many different answers to these questions, some which can be derived from evidence and others that require a political and ethical approach.

Usually, there is plenty of information on each of the aspects mentioned: statistical data, surveys, ethnographic studies, scientific studies, clinical trials, and more. One can safely assume that the more variables included in a particular issue, the more complicated it becomes to define or to prioritize an intervention. The mere existence of information does not guarantee that it will be used, as we will see later.

Once the problem or issue at hand is defined, we need to analyze what to do in order to improve the situation. Different ways of defining an issue will result in different approaches in dealing with it. That is why establishing an adequate definition of the problem is so crucial and to a great extent determines the result of the next stage in the policy cycle.

### 15.2.3 The Making or the Formulation of Public Policy

This stage refers to identifying options for solving the problem previously defined. The assessment of which alternatives are considered viable depends on such elements as social acceptance and pressure, available resources, institutional capabilities and even administrative traditions, among others [4]. The formulation of a policy requires the clear identification the following elements:

- The goals and priorities. A clear indication of the desired results of the policy is a crucial element. For example, it's not enough to say that the policy will improve the quality of life of older adults. It has to specify exactly what that means in all aspects to be covered, as well as which aspects will be dealt with first in accordance with the situation in each community, along with what parameters will be used to measure the improvement in the quality of life. For this, the review of good practices in different countries could be helpful.
- The existing alternatives. In order to achieve the goals stated, alternative packages of interventions must be identified and assessed. Research could contribute in a large extent to tackling this aspect. Being aware of similar policies and programs in national and international settings would make it possible to optimize costs and maximize benefits of an intervention, to avoid duplication, and to promote inter-institutional and multidisciplinary work. An example of this is the World Report on Ageing and Health and the Global Strategy and Action Plan on

Ageing and Health 2016–2020, both published by the WHO. Integrated strategies such as those suggested by WHO allow the organization of different interventions taking into account social and economic outcomes, both in terms of health and wellbeing of older people, along with enabling their on-going participation in society.

- The risks. All interventions may cause unwanted effects, meaning that they could generate situations that could be worse than the alternative of not doing anything at all. Programs could be designed without enough evidence and over time that their benefits are limited and that they even have adverse effects. One such case is the use of vitamin E, which in some places is recommended on a daily basis as an "anti-oxidant"; however, recently, meta-analysis shows that even though it could prevent cardiovascular illnesses and cancer, in high doses it could increase the risk of death [5].
- Alternatives that maximize results. Because of the need to optimize health care expenditures, being able to ensure that an intervention is viable and positive in terms of cost-effectiveness often becomes a decisive factor. In general, this type of study is carried out by both, medical and administrative personnel. Ideally, they do it together since this will allow for the overseeing of the optimization of resources *per se*, but also will also allow for achieving health objectives with more benefits than costs. For instance, healthy aging is the result of lifestyles throughout life. It is thus important to invest in promoting a healthy lifestyle and preventing illness during the course of life.

Ideally, the identification of these elements should be included in the research on age and aging.

Without evidence, the alternatives could be biased by the personal preferences of the decision-makers or be too inertial, just marginally different from past interventions, causing substantial contribution to be made. Even worse, the lack of research or omitting the use of existing research could lead to a worse situation for those affected by policy decisions.

### 15.2.4 The Decision or Choosing Among Options

There are three aspects to be taken into account here: technical viability, which generally considers economic resources, infrastructure and human capacities; political viability, which relates to the probability of cooperation among different actors (public, social and private), as well as leaders of opinion and the media; and social viability, meaning, it is necessary for communities to accept and adopt the policy and become active agents in moving forward with it.

In the case at hand, policies on population aging spread significantly around the world since the 1980s, with the celebration of the World Assembly on Ageing (Vienna 1982). Consensus was clear regarding the impact on development that the increase in the population aged 60 or 65 years and over would have. In this sense, political viability was increased, and government and academia spoke in favor of the aging agenda, resulting in greater political viability. This was followed by an increase in the social acceptability of the issue, thanks to the emergence of a larger and strengthened set of non-governmental actions, increasing the social viability of aging policies.

The third aspect, referring to technical viability, has been harder to achieve, since the generation of specialized infrastructure and human capital compete for resources with other needs that are often considered as higher priorities or that are more profitable politically, such as public safety, gender equality, child protection or the sustainability agenda, to name a few examples.

In cases in which a minimum of political, social or technical viability is not achieved, the intervention should not necessarily be scrapped. It means only that the probability of it being carried fully forward in the long-term is reduced.

### *15.2.5 Implementation*

Once a policy is decided on, we proceed to its execution. For many years this stage was not considered problematic, since there was an instrumental view of public sector organizations that would execute mandates without question or conflict. Bureaucracies exist to implement the orders of their political masters, and disciplinary procedures make sure of this. Since 1970, and as a result of uneven results delivered by well-intentioned policies, this assumption came into question. The implementation can also be a source of innovation and legitimacy of policies. On one hand, organizational capacities and politics can derail a well-designed policy (implementation failure). It could be that the policy itself was faulty (design failure), and public managers are forced to introduce changes in its operation to salvage it and improve its effectiveness and its acceptance by the community. The role of street-level bureaucrats and professionals is of special importance. They have direct contact with the target groups and their skills, political savvy and adaptive capacities can make things work or fail [6]. For example, social workers' decisions as to who is entitled to receive treatment could be optimal or produce errors, such as including people that do not have the intended profile (excess error) or excluding others that do (failure error) [7]. In general terms, it can be said that the implementation of public policy on population aging has been slow, mainly due to a lack of adequate and sufficient infrastructure. The slow development of human resources destined to the specific attention of the older adults, for example, and the scarce formation of geriatricians, have also been important obstacles. However, it is also important to point out that the way we view this issue has been transformed. While in the eighties there was talk of the population of 60 years, it was regarded as just a minority within the general population, while today the impacts of aging on several aspects of human development are recognized.

The implementation of public policy faces a double challenge. On one hand is the need to address the specific needs of older people, taking into account that it is a very diverse group. The needs of older people are different depending on whether they are 60, 70 or 80, if they live in the city or in a rural area, if they are male or female, if they have chronic diseases, etc. There are different approaches to dealing with the problems of an aging population, a population that will face a diverse set of challenges. Promotion, prevention and attention each require a different type of expertise, specialists and infrastructure. On the other hand, finding needed and sufficient evidence to support the aging population requires public policies with a transversal approach, similar to the gender approach, due to the impact it has in all areas of a country's development.

### *15.2.6 Evaluation*

The evaluation of a public policy is, in general, a systematic process of collection and analysis of information for the purpose of showing the efficiency and effectiveness of interventions. Evaluations are a type of research and, therefore, require following technical and ethical standards that apply to research. The evaluation can be internal, that is, to assess the way the agency responsible for the policy is performing, or external, that is, to see if the interventions are generating public value. In the first case, the focus is on the adequate and cost-efficient use of resources (efficiency), as well as on determining the convergence between the objectives programmed *vis-a-vis* the results achieved (efficacy). In the second, the main interest is to identify and measure the effects produced by the intervention (effectiveness).

Evaluations could be classified according to who carries them out, their purpose, their content and the time they are applied. According to Osuna and Márquez [8], an evaluation could be performed by the same team that designs and implements the policy, by an external team, or by a mixed team of insiders and outsiders. Regardless of who performs the evaluation, the most important thing is to define the objective of the evaluation – whether it is for generating information or whether it is required for administrative control, to introduce reforms, to rationalize resources, or to document the lessons derived from the intervention.

It is desirable that evaluation be considered from the beginning of the designing of any intervention. Several aspects can be evaluated: from the design of the policy in order to have a reference of the expected results and the adequacy of the assumed theory of change, to the short-term results and long-term impacts or changes in society derived from the execution of the policy. The implementation and management processes can also be assessed. The evaluation of policy can take place at different moments, providing information on different aspects of the intervention. An evaluation performed after the intervention (ex-post) will produce information about its results and even impacts, given more time and an adequate operationalization of the expected changes in the situations of the individuals that experience a need or a problem. What impact is expected from a low salt diet? What impact would regular physical activity have on a person with diabetes? What would be the results of promoting ballroom dancing among older adults? What would be the indicators and standards for measurement?

An intermediate evaluation or one that takes place during the course of an intervention is focused on gathering details on implementation and management in terms of cost-benefit. Lastly, an evaluation that is performed before the intervention enables us to learn the details of the design of the entire intervention to thereby allow us to identify the coherence between the selected means and the desired goals.

Although many public policies are well designed and implemented, they may not necessarily solve the problem posed, and could even have unwanted effects, or turn out to be excessively expensive and therefore unsustainable. Evaluation has gradually become more relevant to the public policy process. However, it is still most commonly considered only at the end of the period of intervention and implementation, mainly focusing on determining their impacts. The paradox is that, since no adequate methodologies have been established to allow for this sort of evaluation, it is impossible to determine *ex-post* the effects of the policy. It's rare for sufficient economic and human resources to be assigned and the results are not always used to change or even make a decision on whether to finalize the intervention.

There is no doubt that this is an area where scientific research could bring great added value to the process of public policy. Adequate and non-biased evaluation can provide important input for decision makers, providing arguments to defend good policies and increasing the level of legitimacy of public interventions [8].

In the case of the SAGE provides key information on demographic characteristics (age, sex, marital status and education), family arrangements and transfers, participation in the labor force and sources of income, and also, the state of physical and mental health of older adults. This can be used as a formative evaluation that can improve the implementation of public policies in the region. However, it is necessary to follow up on the application of this survey, constantly update information and incorporate different countries into the study. This would make it possible to have data not only for each of the countries, but also for the region, which in turn would allow for interconnection with other initiatives such as the Global Strategy and Action Plan on Ageing and Health 2016–2020.

## 15.3 The Role of Researchers in Decision Making

In the previous section, an overview of the policy cycle has been presented, with emphasis on the potential benefits of using research in each stage. In the following pages the usefulness of research will be further explored along with the reasons policy decisions find using research challenging.

In general, research can contribute by clarifying concepts and allowing to place an issue within a larger theoretical and evidence field. It also can provide tools for making estimations regarding the efficacy, efficiency and legitimacy of different alternatives. It can provide evidence on what has worked in the past in a given policy area. Finally, it can provide information on the potential problems that may have to

**Table 15.1** Contributions of research to the policy process

| Needs of the decision-maker | Potential contributions of research |
|---|---|
| To define public problems | Definition of issues |
| | Determining causes and the components of the issue |
| | Determining who is affected by the issue |
| | Estimating the consequences of different alternatives |
| To formulate alternatives | Alternatives for tackling the problem |
| | Defining objectives and priorities |
| | Estimating risks, costs and benefits |
| | Estimating the monetary cost of interventions |
| To select an alternative | Estimating technical viability |
| | Estimating political viability (social research) |
| | Estimating social viability (social research) |
| Implementation/intervention | Actors who should be involved |
| | Means of ensuring the results |
| Evaluation of the intervention | Criteria for quantitative or qualitative evaluation |
| | Recommendations for improving or terminating an intervention |

be faced and present evidence on the effectiveness of the adopted policies. Table 15.1 summarizes these contributions.

From the previous review of contributions, it is easy to make a case for promoting a closer relationship between researchers from a wide array of disciplines and policy decision–makers. The development of policy sciences was the result of the discipline's founders who sought to create an alliance between science and government [9]. However, achieving this has proven challenging. In fact, it is more the exception than the rule. What are the reasons behind this? For Lindblom and Cohen the explanation is that social science is only one of many ways in which policymakers are informed, and sometimes it is not even the best [10]. Experience and intuition play a role that can be very important. Others blame the misunderstandings between the community of decision-makers and the researchers. Researchers generate information that is of limited relevance to the practitioners or make unrealistic suggestions that do not take into account the restrictions that the public-sector faces. This "two- communities" explanation rests on pointing out that academics and decision-makers live in separate realities, defined by divergent values, incentives and languages [10]. Others have considered this position simplistic: there is evidence of important networks that put practitioners and researchers in touch with each other along with the fact that a sector of public managers, even though not the majority, do make continuous use of scientific evidence [11].

Use requires two elements: that the information is understood and taken into account in the decision-making process and that this process would have been different in absence of such scientific data [12]. There is a wide array of forms for the use of the scientific information. In the case of the results found in evaluations, Landry, Amara y Lamaru [13] define the following steps in a ladder of utilization:

- Transmission: capacity to communicate data or knowledge;

- Cognition: the information is read and understood by the decision-maker;
- Reference: the information is quoted in documents such as reports, plans, programs, etc.
- Effort: there is an effort to adopt the conclusions derived from the information by decision-makers;
- Influence: the information influences the defined alternatives and the decisions made;
- Application: information leads to new uses or new data and knowledge-generating projects.

There are some challenges or obstacles that make it difficult to make use of the evidence found on the higher steps of the ladder. Some are related to the research itself, the clarity of its language and the relevance of the guiding questions, methods and conclusions. Here, the credibility of the source of evidence is also an important aspect. If decision-makers cannot understand the evidence or the relevance of the work, it is improbable that they will use it. Other challenges have to do with to the degree in which decision-makers and other stakeholders are involved in the research process, considering their need for information and managing their expectations. Also, structures and dynamics of organizations can make it difficult to implement the recommendations derived from research even if there are individual practitioners that are convinced of their importance. Lack of resources or institutional capabilities is among the most important organizational barriers for the use of evidence. Finally, individual factors, such as cognitive bias, or competence in data analysis and attitudes towards evidence, can have a significant impact on the probability of evidence being used.

In order to bridge the gap between the supply of scientific evidence and the demand for and the use of the same evidence, we have several strategies or alternatives. Some of the most important are:

- Knowing and understanding the agendas of research and decision-making. The time needed for a research project to bear fruits can find itself in contrast with the immediacy of policy decision-making. Matching the agendas can facilitate decisions in which part of the public policy process will participate. Public officials can also promote and finance research relevant to improving their decision-making.
- The use of language. The researcher must consider using simple and accessible language that can be understood by key decision-makers. Professionalization and adequate professional development interventions can improve the capabilities of public sector managers to understand and use research.
- Creating more teams made up of researchers and decision-makers. The increased interaction between researchers and policy-makers can be beneficial for both sides, as they can develop rapport and a better understanding of each other's needs. An example is the initiative of the *Nesta Operating Company*, (http://www.nesta.org.uk/). This foundation for innovation is located in the United Kingdom, where its objective is to offer investment and subsidies to researchers whose knowledge they believe should be applied and to disseminate that

knowledge permanently thus making technology useful tool [14]. Another example is the creation of policy laboratories (policy labs) where government officials can establish partnerships with experienced researchers to analyze problems, review the effectiveness of present interventions and test new alternatives. The laboratories can also involve a wider set of actors from civilian society in order to crowdsource policy priorities and options and to establish collaborative networks for implementing and monitoring specific programs or projects. There are more than a hundred such initiatives all around the world.

- Maintaining the independence of both the researcher and the decision-maker. There is a need to make sure that researchers are given enough independence to allow them to determine their methods and data-gathering and processing techniques. Ethical standards must be upheld in order to avoid real or perceived conflicts of interest.

Researchers have to be aware of the complexity of public policy. Most problems require more comprehensive and interdisciplinary approaches to be adequately understood. In this case, age and aging are no longer an issue that relates only to the 60-and-over age group. It has become a challenge for all areas of human development. Geriatrics not only includes the study of prevention, diagnosis, treatment and rehabilitation of illnesses related to age, but its goal goes beyond that and involves other disciplines meant to attend to the many social-sanitary aspects that affect the process of aging, illnesses, exposure to risks, long-term care, research, the training of specialists, and more.

Research and generation of knowledge is a public good in its own right. It produces social value when it is translated into better technologies or methods of diagnosing diseases that are increasingly complicated, when it leans toward the generation of models of prevention and replicable care and, no doubt, when it contributes to individuals having better knowledge about their health and how this improves the quality of their lifestyles. Stakeholders should therefore consider knowledge translation as an essential link in the generation not only of public policies but also of public value.

## References

1. Aguilar LF (1991) El estudio de las políticas públicas. Ed. Miguel Ángel Porrúa, México
2. Ripley RB (1985) Stages of the policy process. In: Daniel CM (ed) Public policy theories, models, and concepts: an anthology. Ed. Upper Saddle River, Prentice Hall
3. World Assembly on Aging (1983) A neglected área in the field of population and human rights: ageing and the aged. United Nations, New York, pp 102–109
4. Immergut E (1998) The rules of the game. The logic of policy-making in France, Switzerland, and Sweden. In: Steinmo S, Thelen K, Longstreth F (eds) Structuring politics. Historical institutionalism in comparative analysis. Cambridge University Press, Cambridge
5. Miller ER, Pastor-Barriuso R, Dalal D, Riemersma RA, Appel LJ, Guallar E (2005) Meta-analysis: high-dosage vitamin E supplementation may increase all-cause mortality. Ann Intern Med 142(1):37–46

6. Lipsky M (2010) Street-level bureaucracy, 30th ann. Ed.: dilemmas of the individual in public service. Russell Sage Foundation, New York
7. Cornia GA, Stewart F (1995) Two errors targeting. In: van de Walle D, Nead K (eds) Public spending and the poor. Theory and evidence. The World Bank, Washington, DC
8. Osuna JL, Marquez C (2000) Guía para la evaluación de políticas públcas. Universidad de Sevilla, Sevilla
9. Lasswell HD (1970) The emerging conception of the policy sciences. Policy Sci 1(1):3–14
10. Caplan NA (1979) The two-communities theory and knowledge utilization. Am Behav Sci 22:459–470
11. Newman J (2014) Revisiting the 'two-communities' metaphor of research utilization. Int J Public Sect Manage 27:614–627
12. Leviton LC, Hughes EFX (1981) Research on the utilization of evaluations. A review and synthesis. Eval Rev 5(4):525–548
13. Landry R, Amara N, Lamaru M (2001) Climbing the ladder of research utilization: evidence from social science research. Sci Commun 22:396–422
14. Mulgan G, Puttick R (2003) The case for new institutions. Nesta, UK

# Chapter 16
# Ethical Issues in Research in Aging

**Tirso Zúñiga-Santamaría and Carmen Jimena Vázquez-García**

**Abstract** Medical advancements have resulted in an increment in life longevity, which has led to an increment in the proportion of aging adults in the worldwide population, especially in the last half of the century, as a result of exceeding medical research. This chapter will argue that such exceeding medical research should always be accompanied by an ethical stance, specifically focusing on aging population. The ethical stance in research serves to, first and foremost, look to safeguard the dignity of those it researches. To protect the dignity is not only a matter pertinent to the research process itself, but also to the conclusion of research, and how it is placed within society and the health system. As a response to these important quandaries, the bioethical realm of research has developed methodologies and guidelines for the design and implementation of research protocols to guarantee the protection of dignity and human rights of all and each population studied.

**Keywords** Bioethics · Ethics · Ethics in aging research

## 16.1 Introduction

Current bio-demographic research shows that both life expectancy and longevity of the world population has notably increased in the last century. This phenomenon has been a result of medical research and advancement. However, this reality poses important questions for research going forward. It can be said that today's population reality poses a tension between utilitarian and ethical aspects of society [1]. On the one hand, a utilitarian would call upon the maximization of society's overall utility through research, which we seem to be complying too: but, on the other hand, ethics calls upon a more nuanced outlook on medical research. For example, ethics

T. Zúñiga-Santamaría (✉)
Neurogenetics Department, National Institute of Neurology, Mexico City, Mexico
e-mail: tirzozzu@hotmail.com

C. J. Vázquez-García (✉)
University of Essex, Colchester, UK
e-mail: cv17405@essex.ac.uk

 
C. García-Peña et al. (eds.), *Aging Research - Methodological Issues*,
https://doi.org/10.1007/978-3-319-95387-8_16

would call us to ask: who and how is this research really serving? We have expanded the years in our lives, but have we done the same with their quality? What are the social implications that our medical research has had on the aging? These questions point towards the fact that aging research is not without its ethical complications and, in fact, ethics should be a guiding point for research.

As a first ethical guiding point Baars [2] calls for aging research to always carry within the finitude life and the characteristic that being aware of such fact is what makes us human. This aspect is usually brushed aside because it demerits the idea of utility based on productivity that guides our society on a day to day basis [2]. However, to bring the reality of finitude to the fore, helps achieve a more meaningful and realistic understanding of the aging process, allowing the researcher to see old age as an ongoing learning process of how to confront such undeniable fact [2].

A manner which may result useful into bringing about a positive vision on old age might be to look into history. There are testimonies of previous times when old age was granted as the maximum authority in political, social and cultural aspects. Old age represented the memory of the community, its source of tradition and, therefore, the identity of the people. However, it must also be pointed out that, because their value depended deeply on their memory, if this was lost, they would cease to play an important role in society [3].

*De senectute*, by Cicero, is one of the first Latin works dedicated exclusively to old age; while it is not necessarily a medical book, it incorporates comments related to health and emphasizes the author's view on the matter. In the text there is an argument in favor of old age, representing advantages and possibilities for happiness and well-being [4]. This work by Cicero's holds particular importance to argue against the opinion of many physicians that deem old age as a synonym of diseases, which undeniable implies viewing such life process as an impediment to a good quality of life.

Another important work is that of Laín Entralgo, The Company of Aging, in which he does an analysis of old age by an anthropological, biological, and philosophical approaches. The analysis works by a juxtaposition of the idea of old age against different conceptions of society and different historical moments. In this manner Laín Entralgo is able to show that old age and social life is one of the most serious and urgent issues in the twenty-first century [5].

This chapter will thus, aim to show how aging research has an ethical responsibility to society, specifically to the aging population. The ethical responsibility is not only meant in the safeguarding of dignity for the person over 60 years old taking part in research, but also in the sense that aging research must ethically contribute to a better positioning of the idea of old age in society.

## 16.2 The Problem of Ethics in Aging Research

All research requires taking into account an ethical stance, especially when it deals with a particularly vulnerable part of the population. The ethical dimension of research should be included as a fundamental part in the training of graduate students in the disciplines of health, and should provide logical, epistemological, methodological, philosophical, and ethical statements to researchers [6].

The function of ethics within clinical and epidemiological research is to guide the research without forgetting the dignity of the population being treated. Ethics will not clearly solve the problems that researchers face, but it should always lead the researcher to pause in every step and ask herself: Why am I doing this research? What is my purpose? What will the implication of my research be? The aim of ethics in research is, therefore, to improve the quality of decision making in every step by always having in mind dignity of those participants studied [7]. Within this ethical realm, the duty of health researchers (both applied and basic) is guided by the following criteria: to offer a dignified conception of old age. Such duty is one that covers all aspects of research (whether that be clinical trials, or merely observational), it includes also the very aspect of the motivation for the research itself, and it includes the interaction between researcher and any active participation individual. In short, the ethical is a pertinent and vital part of aging research at all moments and stages of the process.

One of the ways in which the need for ethics in aging research can be seen in a more tangible manner is through the lens of vulnerability. When referencing the aging population, almost always there will be a mention to their (possibly) vulnerable condition, but what does this really mean? Is it a fair characterization? Vulnerability is defined by the Council for International Organizations of Medical Sciences (CIOMS) [8] as the diminution of a subject's capacity to protect their own interests, which, obviously, puts such a person in a disadvantaged position. Such a decrease in the capacity for protection of rights might be due to mental and physical limitations, which older adults have a higher probability of experiencing. This statement is not meant in a pejorative manner, or as a disqualification of old age, but as a stating of one of the possible characteristics of such age. Indeed, human beings can be vulnerable at any time of their existence because of a certain illness or injure, of due to particular circumstances of each if their lives. However, the conditions for vulnerability are usually more prominent in old age which situates the individual at disadvantage regarding the way they are conceptualized in society.

Vulnerability in old age is usually tied up with dependence: insofar aging adults become dependent they become vulnerable. Dependence and vulnerability lead for the people that undergo old age to be treated and understood as different from any other person. Aging adults become the "they" of society, different from the rest because of the negative understanding we have of their life process (one which we will undergo!). This negative understanding undoubtedly impacts the aging's adult quality of life, even more so when such individual is need of care that is granted to them by someone who shares such a negative view of old age. Old age, thus, means—in most cases—a loss of autonomy. The quality of life that the individual over 60 years of age may expect will be, with no doubt, influenced by the way in which society understands and views such a moment in life. Therefore, the individual needs the protection of society and the existence of specialized health-care personnel [9].

The vulnerability of individuals, thus, has consequences for research. One of the most straightforward ways in which we can observe is this is in the repercussion that vulnerability has for the aspect of consent in a research project. As it has been said, one of the ways in which an aging adult is come to be seen as vulnerable, has to do with limited cognitive or physical ability, which may impact her capacity to make judgments and decisions. This is especially worrying when persons over 60 years of age are invited to participate in research protocol in which they can be coerced into participation by a family member, or by the researcher itself, particularly when legal competence has not been clearly established. And just because coercion does not appear evident, it does not mean it might not be there. For example, aging adults might feel pressured to agree on participation because they deem the researcher, or family member, as a figure of authority that they cannot deny. Or, it can also be the case, where the individual agrees to participate because they find comfort in the attention, and in cases as these, the researcher must be very careful not to take advantage of this situation.

However, the relation between vulnerability of subjects and research is more complex than that. Assuring that the individual has granted full and informed consent to participate in a research project is obviously important but vulnerability may play a bigger role than that. Vulnerability is not only at play in clinical trials but also in observational studies. For example, there could be a case in which a researcher is "merely" asking questions to the aging adult, and such questions may leave the individual distraught. Or, it may be the case in which during the observation, the researcher notices that the individual in questions suffers from certain conditions that she is not aware, and the researcher does not follow up to assure a suitable treatment. All of these are instances in which the role of vulnerability may not have been as evident as in the process of granting consent, but it was nonetheless there. It may be the case that the individual was not vulnerable from the start but suffered vulnerability at some point in the research.

Another matter which needs to be address is that the importance of research done ethically is not limited to the process of research itself and the way it might endanger such a vulnerable population but is also implicated in the connotations it serves to create in society regarding old age. A simple example of this is the common use of certain labels within research that are used to refer to the aging adults, like frailty. Even if the researcher does not make use of such a term with an intention of a negative connotation, she must take into account that the way which her research is done might lead to people understanding such a term in negative ways and, thus, have an impact on the individual's life; because as it was just stated the quality of life of the person depends on the quality of care she receives, which will be connected to the value the care giver gives the life she is caring for. However, the use of such labels may also even impact aspects within the research itself such as the selection process in which certain persons of old age may be discriminated against. Even the selection process, hence, has clear bioethical implications for researchers and health professionals [10].

Because of the implications pointed out, it is clear that as difficult as it may be, there must be a careful and reflective analysis of the values and principles in conflict during every stage of research: the decision making cannot consist of a pure mathematical equation [6].

As Weber pointed out, we must consider our ethical responsibility in the consequences of our actions: when we make decisions, we must also asses the consequences and assume responsibility [11]. We must therefore asses the consequences of every step of the research process, even of such preliminary steps as the decisions made regarding how we refer to the individuals whom we are studying. We must accept that the use of word "elderly" over "individual" will have an impact, whether that was our intention or not, and different connotations and judgments will arise. We must ask ourselves if through our research we are ensuring the respect of those studied. We must never lose sight of whom the individual is, we must never cease to recognize her as human, and our responsibility towards her. This is a vital part of our responsibility and obligations as researchers. David Oliver notes that the only way to avoid getting old is to stop living [12]. In other words, to get old is an undeniable part of human life for the privileged ones that made it so far. Why then, do we treat old age, as something wrong and undesirable?

## 16.3 Ethical Considerations in Aging Research

Having argued in favor of an ethical realm within ageing research, it is now prudent to discuss the considerations that such realm imposes on research. In "Bioethics as a discipline" (1973) Daniel Callahan [13] foresees the appearance of a discipline called bioethics and reflects on the role that ethics could take in the world of health and biology. Callahan states that philosophers must be a part of health and biology research in order to offer intellectual rigor to the problems that arise within such instance; for example, to help health professionals, researchers and biologists in making concrete decisions. Bioethics, thus, serves to define ethical problems, methodological strategies and procedures for making decisions that are sensitive to the complex cases. This shows that bioethics is interested in the adequate application of the ethical statements and ethical dilemmas; it is more concerned with solving ethical dilemmas than with the formation of character and the search for excellence in the health professional and in the researcher [14].

However, even bioethical research seems as relatively new development, the reality is that in some way or another, bioethical aspects of research regarding humans have always been part of research since the beginning of Medicine. However, it was not until the second half of the twentieth century that bioethics took a more prominent role, due to very disturbing and worrying historical events, that enforced a more thorough evaluation of research protocols. The example that undoubtedly comes to mind is the way that "medical research" was performed by the Nazis. Such atrocities gave rise to the Nuremberg Code (published 1947), which served not only to install an undeniable ethical base for future research, but also served to judge the researchers during the Nuremberg Trial processes [15].

Another case that is important to mention is that of Jewish Chronic Disease Hospital [16], in New York (1963), where cancer cells were injected into 22 patients of old age without their consent. The "justification" rationale given by the researchers was that they sought to discover whether in cancer patients there was a decrease in the body's ability to reject transplants with cancer cells because of weakness or the aging process. Of the 22 patients, none were informed of what was being done to them, and only some knew they were a part of an ongoing medical research.

These are just some of the cases that led Henry K. Beecher to work upon the distinction between the therapeutic and the non-therapeutic experiments, which in turn resulted in the Declaration of the World Medical Association, signed in 1964 in Helsinki. Such document, of which Beecher was one of the main drafters, remains valid to this day, even if it has undergone several successive reforms. In its most recent version (2013) [17] it states: "Medical research is subject to ethical standards that serve to promote respect for all human beings and to protect their health and their fundamental and individual rights".

Following such declaration, Beecher published an article titled "Ethics and Clinical Research", where he denounced investigations with serious ethical problems. The main problems to which he signaled to where: informed consent, the evaluation of risk-benefit (in very high risk), and the fact that the experiments involved vulnerable populations (children and the persons over 60 years of age). Upon this, Beecher concluded: "An experiment is or is not ethical from the beginning; it does not become ethical post hoc, the end does not justify the means" [18]. Beecher's article showed, that even if Declarations and Codes are drafted, they are not necessarily enforced in research. One of the manners in which this has been tried to solve is by posing more importance into the aspect of informed consent as a fundamental part of the evaluation of research protocols.

Informed consent, it has been established, is an important ethical condition of research as it signifies that respect for the subject's autonomy is being taken into account. Hence any participant of research must always be informed in a clear and concrete manner about what their participation in the research entails. To highlight the importance of informed consent in research is to safeguard the dignity of vulnerable parts of the populations, which as we have seen is a characteristic of old age. This is why such a parameter is one of the outmost important conditions imposed by the ethical real in research. The old age population has a greater probability of living with conditions that limit them physically and mentally, which situates them in a position of complete disadvantage use of their diseases and might not be able to grant or reject informed consent themselves and are vulnerable to influence from the researchers.

Ethics in research has, thus, resulted in methodologies and regulatory rules for clinical trials in order to protect the human rights of research subjects. These normative measures, though existent in some form or another throughout history, were expressly written out from the aftermath that followed the horrific historic events of the mid twentieth century. Such guidelines should be seen a living organism always to be analyzed and allowed to evolve, in order to truly protect humans and in order to meet the debates of the current times, like those of embryonic stem-cell research or euthanasia.

## 16.4 Research Ethics Committees for Aging Research

In order to better guarantee the ethical conditions that research must meet, institutions that do research on humans must have a Research Ethics Committees (REC) that regulates the research protocols. A REC is made up of a multidisciplinary group of people within a health institution (independent of the government system), and such institution may conduct research on both animals and humans. Such group of people should include researchers from different fields of scientific knowledge, both from the institution itself and external to it and it should also include members of the community [19].

The fundamental objective of the REC is to safeguard the human rights and dignity of the research subjects. This implies knowing and evaluating the probable benefits risks of the research for the participating subject before the research is performed and evaluated. Both benefits and risks should be clearly stated in the research protocol, as well as other ethical aspects, such as the financing of the research and the relationship of the principal investigator with the pharmaceutical industry, among others. However, as it can be inferred, REC is not alone in ensuring ethics in research. Currently there are other important set of norms and directives that also set guidelines to ensure the respect of human rights within research: UNAIDS, 2000; CIOMS, 2002; Nuffield, 2002 [20, 21].

Without a doubt the knowledge generated by research on humans is necessary and valuable, it has allowed the development of medicines and vaccines, for example. But these advancements in themselves are not enough to justify an undignifies treatment of research subject, which is why all research must be first and foremost scientifically and ethically adequate, especially in the case of the vulnerable parts of the population. Herein rests the importance of the REC.

## 16.5 Discrimination of the Problem in Aging Research

To safeguard the dignity of individuals can be also be phrased as to protect them versus discrimination. All human beings have the right to live their lives fully and in a meaningful way. Aging research should, therefore, have such a right of human life as the ultimate objective for any medical advancement. Nonetheless, as this chapter has been trying to show, certain aspects of aging research have generated negative connotation in the way that society sees old age which has cause such individuals to be seen as less worthy of a good life. One of the manners in which aging research, purposely or not, has negatively impacted the lives of the aging adults, is by the creation and subsequent use of certain labels. One of the labels we have already exemplified is that of frail.

The term frail references an accumulation of deficits which signify a distinct clinical syndrome insofar it involves a specific co-morbidity, such as physical disorder and functional impairment. Frailty has become an increasingly common component of aging research insofar the probability to become frail in old age is higher. A fundamental goal of geriatric medicine has, therefore, become how to maximize functionality, independence, and quality of life during the years lived with frailty for as long as possible [22].

Nonetheless, the reality is that the concept of frailty is still under development and analysis within aging research; it has "little robustness" and is in need of more scientific evidence to validate it and its transcendence. Even more so, more research is needed specifically regarding frailty's clinical aspects and unique psychological aspects, so as to be able to accurately define, recognize, and treat frail individuals. Therefore, research regarding old age should not just scientifically enhance the concept of frailty, but also help to construct a positive narrative regarding such condition. As of today, frailty is quickly equated with many negative aspects which ultimately cause several types of discriminations towards old age. Thus, any research on aging should be careful to not imply that frailty lessens the person's life value to society.

Even if the researcher does not make use of such a term with an intention of a negative connotation, she must take into account that the way which her research is done might lead to people understanding such a term in negative ways and, thus, have an impact on the person's life; because as it was just stated the quality of life of the person depends on the quality of care she receives, which will be connected to the value the care giver gives the life she is caring for. However, the use of such labels may also even impact aspects within the research itself such as the selection process in which certain persons of old age may be discriminated against. Even the selection process, hence, has clear bioethical implications for researchers and health professionals [10].

## 16.6 Aging Research, Ethics and Health Systems

Discrimination, as the previous section argued, does not arise out of thin air, but as a result from human action. There is a legal and a moral dimension to discrimination: a fundamental human right is being violated, and such violation is considered unfair. And in the same manner that discrimination is enacted by humans, so is morality and our capacity to become ethical in our going about life. Our capacity to be moral and ethical is particularly important when analyzing the ends, we pursue, because it is in such a manner that the responsibility we hold with our actions becomes truly evident. Our actions, thus, should not be concealed by the way things are, but by the ideal of the way things should be. What this chapter has tried to outline is that this is true also in aspect of health and aging research. In this last section, we will know try to show that aging research done ethically could also help health systems in avoiding discrimination.

Ethical guidelines, for example, could help set up strategies regarding the allocation of economic and health resources. In this sense, and ethical outlook, could help with difficult decisions where the most prudent allocation might not seem the fairest at first sight, a matter which is not talked about enough and endangers the optimal distribution [23].

Another aspect in which the health systems could gain from an ethical outlook is regarding specifically to policy aspects. The policy decision, in many countries, to shift the burden of health care and social security for the persons over 60 years of age from families to the State has had major socio-economic consequences. Moreover, the economic pressure on families has not been alleviated, especially in the cases where the person is in need for long term care, whether that be institutional or home based. Needless to say, that as the proportion of the aging adult increases so does, the economic pressure on the State [1]. These difficulties lead us to ask more in-depth questions regarding the role of the State in such instances: where is the line drawn in the State's obligation to the older individuals? What should the scope of public health systems be in regard to care? What is the interplay that should be expected between health research and State to be? In other words, should the State have a say in the type of research being done? Should it guide the prioritizing of certain parts of the population over others as research objectives?

A manner, in which the previous questions could be tackled, would be by a sort of ethical principle to accompany the decision making of health systems. In this sense Norman Daniels has argued in favor of a principle that can be applied, both on the matters of the need for aging research and the urgency of care: "[...] meeting the needs of health care is a matter of special importance because it favors equal opportunities. It serves to guarantee individuals a fair option to enjoy the normal range of possibilities of the society in which they live" [24]. Daniels claims, thus, that such code that should lead policy decision making should be one that favors equal opportunities as such condition will work in favor of guaranteeing quality of life.

To point towards this difficult relation between aging and health systems is not meant as a demerit of the achievement from modern health and science, which has signified the rise of the aging population. It is merely a signaling towards the aspects that we should further problematize and analyze in order to grant the possibility for persons over a certain age to continue to be a driving force in society; a matter which depends on our ability to safeguard their dignity [7].

## 16.7 Conclusions

Ethics is as old as Medicine. However, even though in theory ethics and medicine should be found to go hand in hand, this has not always been the case. Horrific historic instances have showed how vulnerable the human being can be in the name of medical advancement; these moments lead to a better questioning of how to truly achieve and safeguard ethical standards in research. Declarations and rights have been written and set out with the purpose of protecting the dignity of those who participate in medical research, especially in the case of vulnerable parts of the population.

A particularly vulnerable part of the population is the aging adult, and so, this chapter has also tried to reflect how research can also signify a risk for old age if not done by ethical terms. Given the particular vulnerability of the aging population (frailty, dependence, disability) it is of the outmost importance that ethical standards guarantee the voluntary participation of such individuals in any research and that their dignity is safeguarded all along the process. However, the importance of ethics in aging research does not end there. Any researcher should also be concerned about the consequences of their research and the connotations which it will have in society. Therefore, aging research should pay particular attention to the fact that their research should serve, overall, to better the quality of life of all aging adults with regards to their place in society. In other words, all aging research should treat old age with the respect that they expect society to do so. If any aging research is serving towards the advancement of medicine but feeding into the palpable and worrying discrimination of the old because of negative connotations, we have to ask ourselves, as researchers, how valuable such research actually is.

## References


1. Callahan D (2004) Poner límites: los fines de la medicina en una sociedad que envejece. Triacastela, Madrid
2. Baars J (2017) Aging: learning to live a finite life. Gerontologist 57:969–976. https://doi.org/10.1093/geront/gnw089
3. Gracia D (1998) Ética de los confines de la vida. Editorial El Búho, Santa Fe de Bogotá
4. Cicerón M (2001) Acerca de la vejez. Triacastela, Madrid
5. Laín-Entralgo P (2001) La empresa de envejecer. Galaxia Gutenberg, Madrid
6. Gracia D (2008) Fundamentos de bioética, 2a edn. Triacastela, Madrid
7. Zúñiga T (2013) Bioética y calidad de vida en ancianos con demencia, 1ra edn. Médica Panamericana, México
8. Council for International Organizations of Medical Sciences (2002) International ethical guidelines for biomedical research involving human subjects. CIOMS, Geneva
9. Zúñiga T (2015) Bioética en la investigación con ancianos. Ensayos sobre ética de la salud: investigación Volumen 2: Aspectos sociales. 1a. ed. Universidad Autónoma Metropolitana, México, pp 83–90
10. Gracia D, Júdez J (2004) Ética en la práctica clínica, 1a edn. Triacastela, Madrid
11. Hans J (1995) El principio de responsabilidad Ensayo de una ética para la civilización tecnológica, 1ra edn. Herder Editorial, Barcelona
12. Oliver D (2017) How can we plan for old age if we won't discuss it honestly? BMJ 357:j2759. https://doi.org/10.1136/BMJ.J2759
13. Callahan D (1997) Bioethics as a discipline. In: Jecker NAS, Jonsen AR, Pearlman RA (eds) Bioethics: an introduction to the history, methods, and practice. Jones and Bartlett Publishers, Massachusetts, pp 87–92
14. Gracia D (1998) Hacia un enfoque socrático de la enseñanza de la Bioética. Fundamentación y enseñanza de la bioética, 1era edn. Editorial El Búho, Santa Fe de Bogotá, pp 45–57

15. Curran WJ (1969) Governmental regulation of the use of human subjects in medical research; the approach of two federal agencies. Daedalus 98:542–594
16. Alliance for Human Research Protection. (1962) Dr. Chester Southam injected live cancer cells into 22 elderly patients n.d. http://ahrp.org/1962-dr-chester-southam-injected-live-cancer-cells-into-22-elderly-patients-at-jewish-chronic-disease-hospital-in-brooklyn/
17. Declaración de Helsinki de la Asociación Médica Mundial (2013) Principios éticos para las investigaciones médicas en seres humanos
18. Beecher HK (1966) Ethics and clinical research. N Engl J Med 274:1354–1360. https://doi.org/10.1056/NEJM196606162742405
19. Zúñiga T (2014) Ética y bioética en las instituciones de salud. In: Malagón LG, Pontón Londoño JR (eds) Auditoría en salud: para una gestión eficiente, 3ra edn. Médica Panamericana, Bogotá, pp 22–8
20. UNAIDS (2000) AIDS: palliative care. UNAIDS Technical update, Geneva
21. Nuffield Council on Bioethics (2002) The ethics of research related to healthcare in developing countries. Nuffield Council on Bioethics, London
22. Fillit H, Butler RN (2009) The frailty identity crisis. J Am Geriatr Soc 57:348–352. https://doi.org/10.1111/j.1532-5415.2008.02104.x
23. Gracia D (2004) Ética de la no-discriminación. In: Médicos (ed) Clínicas Geriátricas. Editores Médicos, Madrid, pp 259–269
24. Daniels N (1985) Family responsibility initiatives and justice between age groups. Law, Med Health Care 13:153–159. https://doi.org/10.1111/j.1748-720X.1985.tb00911.x

# Chapter 17
# Integration of Consortiums and Search for International Funding

**David X. Marquez, Iraida V. Carrion, Susan Aguiñaga, and Melissa Lamar**

**Abstract** There is a rising number of older adults in the United States, and around the world. There are health, economical, and social implications and decisions to be made that affect all of us, including older adults. Data need to be gathered in order to inform these important decisions. Thus, there is a critical need for (1) societies to develop greater investments in aging-related research, (2) researchers to apply for funding in order to develop evidence-based practice and programs to prevent or treat the major causes of diseases and disability, and (3) recruitment and retention of older adults in research. In this chapter we describe funding mechanisms related to aging research, given that funding allows researchers to generate data and create novel interventions. We describe the process involved with searching for aging research funding and discuss important considerations when writing aging-related grant proposals. Additionally, we describe the importance of and methods on, forming international collaborations.

**Keywords** Grant writing · Funding · International consortiums

---

D. X. Marquez (✉)
Department of Kinesiology and Nutrition, University of Illinois at Chicago, Chicago, IL, USA

Rush Alzheimer's Disease Center, Chicago, IL, USA
e-mail: marquezd@uic.edu

I. V. Carrion
School of Social Work, University of South Florida, Tampa, FL, USA
e-mail: icarrion@usf.edu

S. Aguiñaga
Department of Kinesiology and Public Health, University of Illinois at Urbana-Champaign, Champaign, IL, USA
e-mail: saguina2@gmail.com

M. Lamar
Rush Alzheimer's Disease Center, Department of Neurological Sciences, Rush Medical College, Chicago, IL, USA
e-mail: Melissa_Lamar@rush.edu



C. García-Peña et al. (eds.), *Aging Research - Methodological Issues*,
https://doi.org/10.1007/978-3-319-95387-8_17

## 17.1 Introduction

The twenty-first century is experiencing a significant growth in the older adult population. For example, in the United States older adults aged 65 years and older grew from 35.0 million in 2000, to 49.2 million in 2016 [1]. Worldwide, the number of older adults aged 60 years and over is projected to grow by 56% between 2015 and 2030 [2]. Advances in public health and technology, reduction in child mortality, and rising living standards have contributed to people living longer than ever before. Although this demographic change reflects advances in socioeconomic development and progress in human development, it is not without challenges. Older adults' longer lives are often fraught by sequelae of diseases and injuries.

Chronic non-communicable diseases (e.g., cardiovascular disease, cancer, chronic respiratory disease, and others) account for most of the global disease burden in older adults [3]. Older adults have the highest prevalence of multiple chronic conditions of which 77% of older adults have at least two [4]. The increase in chronic diseases is due to a shift in risk behaviors such as diets low in fruits and vegetables and high in sodium and saturated fats, tobacco and alcohol use, and physical inactivity [3]. In order to reduce the chronic disease burden, interventions at the individual and population levels, as well as epidemiology and surveillance to monitor trends and track progress need to continue to be developed and implemented [5].

The prevalence of neurodegenerative diseases, including Alzheimer's disease and other dementia-related diseases has also increased, as older age is associated with these diseases. In 2010, Alzheimer's disease and other dementias affected 35.6 million worldwide. It is projected that this number will increase to 66 million by 2030 [6]. As a global society, we are currently not prepared for the provision of care that such an increase will incur on our health systems and economy. Furthermore, Alzheimer's disease is the only cause of death that cannot be prevented or cured, and, as such, needs to be regarded as a global health priority.

The unprecedented demographic change that the world is encountering will lead to economic and social shifts. These changes are galvanizing policy makers, clinicians, and researchers worldwide to improve the care of older adults, maintain their quality of life, and protect against age-related decline. In order to tackle the issues that come with the negative effects of aging, data need to be derived in order to inform these important decisions. Thus, in order to derive data, there is a critical need for (1) societies to develop greater investments in aging-related research, (2) researchers to apply for funding in order to develop evidence-based practice and programs to prevent or treat the major causes of diseases and disability, and (3) recruitment and retention of older adults in research. Furthermore, there is an urgent need for research to be conducted among older adults in low-resource countries of the world, as low-resource countries are also confronting these demographic changes; however, data are only now becoming available.

In this chapter, we focus on funding mechanisms related to aging research, given that funding allows researchers to generate data and create novel interventions. We describe the process involved with searching for aging research funding and discuss important considerations when writing aging-related grant proposals. Additionally, we describe the importance of and methods on, forming international collaborations.

## 17.2 Searching for Funding in the United States

There are many considerations to be made when searching for funding to apply for. One consideration is the type of funding, federal or foundation. At the federal level within the US, the most common funding source is the National Institutes of Health (NIH), which has strict rules for applying for the grants (https://grants.nih.gov/grants/how-to-apply-application-guide.html and https://www.nsf.gov/funding/preparing/). At the NIH many aging-related proposals are directed to the National Institute on Aging (NIA), one of the 27 Institutes and Centers of NIH that "leads a broad scientific effort to understand the nature of aging and to extend the healthy, active years of life." They state that they are the primary Federal agency supporting and conducting Alzheimer's disease research. However, many of the 27 Institutes of the NIH fund aging-related research, depending on the focus and funding priorities set forth by the institute. For example, research focused on age-related emotional regulation and processing should be directed to the National Institute on Mental Health (NIMH).

There are several types of funding opportunities available from the NIH (https://grants.nih.gov/grants/how-to-apply-application-guide/prepare-to-apply-and-register/understand-funding-opportunities.htm), and various ways to learn more about the mechanisms or particular topic areas of interest to the NIH more specifically. First, Parent Announcements are broad funding opportunity announcements (FOAs) in which applicants submit an application for a specific activity code (e.g., R01, R03, R21 are several of the main types of funding, as outlined below). Many NIH institutes and centers participate in these FOAs, they are usually ongoing, and these R mechanisms use the NIH standard due dates. Second are Program Announcements (PA), which are from one or more Institutes and Centers to highlight areas of scientific interest. There are PAs with set-aside funds (referred to as PAS), and PAs with special receipt, referral, and/or review considerations (referred to as PAR). Third are Request for Applications (RFA), which are also from one or more Institutes or Centers to highlight well-defined areas of scientific interest to accomplish specific program objectives. However, these have the amount of set-aside funds indicated, state the anticipated number of awards to be given, and usually have a single due date not too long after the RFA is posted.

The NIH has different activity codes, depending on the size and objectives of the proposal (https://grants.nih.gov/grants/funding/funding_program.htm). Generally, PIs and other personnel supported by NIH research grants are not required to be US citizens; however, some NIH programs/mechanisms have a citizenship requirement. Any citizenship requirement will be stated in the PA or RFA. The NIH strongly encourages non-US applicants to review the Eligibility section of the FOA to determine whether their non-US organization is eligible to respond to that particular FOA.

The NIH offers numerous research career development awards to support researchers (https://researchtraining.nih.gov/programs/career-development). For example, there is funding to support: early career scientists in need of both advanced research training and additional experience (K01); promising clinician scientists with demonstrated aptitude to develop into independent investigators (K08); experienced scientists who wish to broaden their research capabilities or to make changes in their research careers by acquiring new research skills or knowledge (K18); or to support both an initial mentored research experience (K99) followed by independent research (R00) for highly qualified, postdoctoral researchers, to secure an independent research position.

Some of the more common funding types to apply for include the NIH Research Project Grant Program (R01) which is used to fund a discrete, specified, circumscribed research project; is the most commonly used grant program; has no specific dollar limit with the exception that direct funds for research cannot exceed $500,000 in any given year unless specified in the FOA and/or requested to exceed that value by the Principal Investigator to the Program Official, and is generally awarded for 3–5 years. There are smaller funding opportunities, including the NIH Small Grant Program (R03) which provides limited funding (maximum of $50,000 per year in direct costs, not exceeding $100,000 over 2-years) to support a variety of types of projects, including: pilot or feasibility studies, collection of preliminary data, secondary analysis of existing data, small, self-contained research projects, development of new research technology, etc. An R03 is not renewable. A mechanism with available funding between that of an R01 and R03 is the NIH Exploratory/Developmental Research Grant Award (R21) which solicits new, exploratory and developmental research projects by providing support for the early stages of project development. Like the R03, the R21 is limited to 2 years of funding.

Foundation funding is also available. Some foundations have a general research mission, with specific research objectives. For example, the Retirement Research Foundation (http://www.rrf.org/) has a mission to improve the quality of life for our nation's older adults. They have Responsive grants in the areas of advocacy, direct service, professional education and training, and research; and Organizational Capacity Building (OCB) grants that provide funding for improvements in key management and governance functions within nonprofit organizations that serve older persons in the Chicago area.

Foundations often have a more direct and specific focus for funding. For example, the Alzheimer's Association conducts the Alzheimer's Association International Research Grant Program (https://www.alz.org/research/alzheimers_grants/overview.asp). They state that the impetus for their awards is their "desire to improve quality of life for people affected by Alzheimer's." Clearly, this type of aging-related research is more focused than general aging research. The American Federation for

Aging Research (https://www.afar.org/) is another nonprofit organization and has a mission to support and advance healthy aging through biomedical research. This focus on biomedical research is important to note, as those who are doing research in non-biomedical aging research might look for other funding sources.

The Sociological Initiatives Foundation (http://www.sifoundation.org/) funds research that supports social change by linking research to social action. This type of funding provides initial funding when addressing caregiving and aging in place issues. It funds research projects that investigate laws, policies, institutions, regulations, and normative practices that may limit equality in the United States. It gives priority to projects that seek to address racism, xenophobia, classism, gender bias, exploitation, or the violation of human rights and freedoms.

The McKnight Endowment Fund for Neuroscience (https://neuroscience.mcknight.org/) is an independent charitable organization that supports innovative research focusing on diseases of the brain and behavior, such as Alzheimer's disease, Parkinson's disease, multiple sclerosis, spinal cord injuries, and others. The Dana Foundation (http://dana.org/) is a private philanthropic organization that supports science and health grants through neuroimaging and clinical neuroscience research. In short, foundation funding related to aging is available, however, it is important to know the mission of the foundation and what types of research they prioritize to fund.

Researchers/investigators might also look to pharmaceutical/medical companies for funding. Some universities have restrictions on such funding relative to what their Institutional Review Board will review, or charges for services if the funding is from a pharmaceutical company. That said, they are often less restrictive of international PIs. For example, Abbott Fund (http://www.abbottfund.org/grants) was established by the company Abbott, a global health care company, in 1951, as a philanthropic foundation. They invest in ideas that promote science and medical innovation, and expand access to health care and strengthen communities around the globe. Other corporations like the Baxter Healthcare Corporation (http://www.baxter.com/inside-baxter/science/programs/medical-research-grants.page) provide support for programs that advance scientific research, medical education and patient care.

## 17.3 Important Details when Writing Proposals for Funding for Research on Aging

When writing "aging-related" proposals it is important to consider several factors. First, how is aging defined? It is a biological process? Or is the research more about aging-related factors? Next, what aspects of aging are being targeted? Environmental? Biological? Psychological? Cognitive? Physical? Cellular? Molecular? A combination of these or other factors?

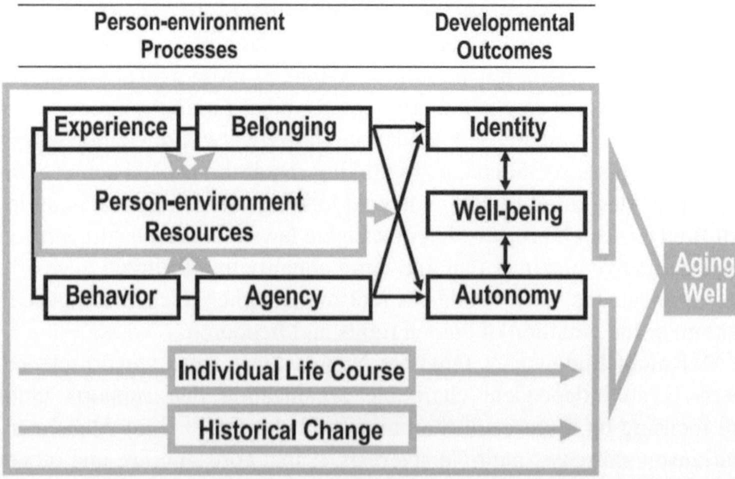


**Fig. 17.1** Conceptual framework: interplay of belonging and agency, aging well, and the environment [7] Licensed by Oxford University, Number: 4293140398195

Another factor to consider is whether the research takes a lifespan approach or focuses on a specific age cohort, like middle-aged or older adults. For example, the objective of the research might be to see how circumstances or changes in midlife influence those of older adults. Thus, the focus is not on older adults, but an earlier age range that may have a lasting impact into older ages. Alternatively, the objective might be to examine the lives of current older adults, to describe, explain, or predict future aging. Research with older adults is frequently used synonymously with aging research, although aging research is not specific to work with people 65 years of age and older. These distinctions are important to consider when thinking of your funding body be it federal or foundation.

Researchers might consider using a framework or model of aging to explain their work and the relationships among variables. For example, Wahl, Iwarsson and Oswald (2012) proposed a conceptual framework of aging and the environment (see Fig. 17.1). The basic assumption of this framework is that two processes, experience-driven belonging and behavior-driven agency, assist in forming a better understanding and integration of existing Person-Environment interchanges *as people age*. This framework has a focus on the dynamic changes that occur with aging and over the lifespan, as people and their environments change. It includes the role of the immediate physical, spatial, and technical environment, which has been largely ignored in gerontological research, and is ever-so important in today's world. The authors state that a major implication of the model for future research is the importance of explicitly considering aging in the environment in longitudinal studies of aging rather than decontextualizing the aging individual, as most longitudinal studies tend to do. Thus, we can see that research focused on aging and the lifespan can benefit from frameworks such as this one.

## 17.4 International Collaborations and Funding

International research collaborations improve the possibility of external funding. Establishing and synchronizing planning takes time, but given communication technology, it is possible and can have many positive outcomes. Many academic institutions, medical centers, and foundations provide pilot study funding opportunities for new researchers as well as for senior researchers. This vehicle for preliminary funding provides an ideal opportunity for international collaboration and future funding. At times, universities will make funding available for a researcher to attend or present preliminary research findings at an international conference or congress, where researchers can also network and establish working relationships with colleagues of similar research interests. Additionally, when applying for foundation grants and international conference grants offered by academic or medical institutions, researchers can include preliminary communications and collaborations that have been established with international colleagues. In the grant application, the visiting researcher can also indicate that they will schedule an in-person meeting with an identified researcher in the country in which the international conference is being held in order to further develop a grant application or a joint publishable paper. It has become increasingly necessary to have published works with colleagues prior to applying for research funding. This allows the visiting researcher to become further acquainted with potential research colleagues and/or research teams working in areas of mutual interest. This also enables the visiting researcher to become familiar with the inner workings of institutions, grant mechanisms, collaborative research opportunities, and community partners. Given that researchers tend to attend the same conferences or congresses annually, they can coordinate enhancing and concretizing their research funding endeavors with these professional meetings.

In order for a visiting researcher to learn about the context of the research setting, it is necessary that they engage with the communities they aim to study. Historically, researchers have excluded community members, leaders, and stakeholders from research endeavors. It is therefore essential that these players be included in the conceptualization phase of international research endeavors, given that community engagement will strengthen and expand proposed research projects.

One way to enhance one's research internationally is through the International Aging Research Portfolio (IARP; http://agingportfolio.org/), an independent joint initiative of government, academic, corporate, patient advocacy, and charitable funding organizations. IARP is a flexible system that enables funding organizations to collaborate, track, analyze, structure, make decisions, and set directions for future research efforts in aging; it also address the needs of research investigators, health care policymakers, government officials, interest groups and the general public. It tracks international progress in aging research and provides a wide array of information on funded projects, including projects funded by the Center for Global

Health. The Grant Match Maker(TM) tool within the IARP is designed to identify prospective sources of project funding, identify similar or related projects that have successfully received financing in the past, and help classify the project into relevant categories. The system uses the project categorization algorithms (see the "Aging Portfolio Classification Algorithms") to identify related projects and categories within the Aging Portfolio database. The use of international research conference grants, seed grants, foundation grants, and IARP as opportunities to create future collaboration partnerships has been successful for researchers. The following website, http://staff.lib.msu.edu/harris23/grants/privint.htm, provides a list of international and foreign grant makers, with links to various funding sources. It is possible to obtain funding with international collaborators, but this requires strategic planning and cultural humility in interactions with members of the international community. Exchange opportunities can enhance and create research funding collaborations, and researchers can gain a global perspective in their efforts to address the needs of older adults.

## 17.5 Conclusions

Funding in the US and around the world is becoming more competitive, often with more investigators applying for reduced funds over time. Given the changing age demographics of most countries, aging-related research is a focus of many governments and foundations. However, there is more work to be done, as funding is not keeping up with age-related problems people are experiencing. By informing oneself of various funding agencies and mechanisms, and working with a strong collaborative team, success can be achieved so we can continue to learn and help improve the lives of older adults.

## References

1. US Census Bureau (2017) The nation's older population is still growing, Census Bureau reports. Available at: https://www.census.gov/newsroom/press-releases/2017/cb17-100.html. Accessed 31 Jan 2018
2. World Population Ageing 2015 (n.d.) Available at: http://www.un.org/en/development/desa/population/publications/pdf/ageing/WPA2015_Report.pdf. Accessed 31 Jan 2018
3. Prince MJ, Wu F, Guo Y, Gutiérrez-Robledo LM, O'Donnell M, Sullivan R, Yusuf S (2015) The burden of disease in older people and implications for health policy and practice. Lancet 385(9967):549–562. https://doi.org/10.1016/S0140-6736(14)61347-7
4. National Council on Aging (2014) Healthy aging fact sheet. Available at: https://www.ncoa.org/wp-content/uploads/FactSheet_HealthyAging.pdf

5. Bauer UE, Briss PA, Goodman RA, Bowman BA (2014) Prevention of chronic diseas in the 21st century: elimination of the leading preventable causes of premature death and disability in the USA. Lancet 384(9937):45–52 https://doi.org/10.16/S01406736(14)60648-6
6. Wortmann M (2012) Dementia: a global health priority- highlights from an ADI and World Health Organization report. Alzheimers Res Ther 4(5):40. https://doi.org/10.1186/alzrt143
7. Wahl HW, Iwarsson S, Oswald F (2012) Aging well and the environment: toward an integrative model and research agenda for the future. Gerontologist 52(3):306–316. https://doi.org/10.1093/geront/gnr154

# Chapter 18
# Future of Aging Research

**Matteo Cesari, Marco Canevelli, and Mario Ulises Pérez-Zepeda**

**Abstract** The world population is aging, and this phenomenon is expected to continue for the next decades. The demographic modifications are substantially altering the socio-economic structure of our societies and pose serious burdens to the sustainability of healthcare systems worldwide. In order to guarantee that models of care are responsive to the novel needs and transformations, it is necessary to invest in new research paradigms that are respectful of the complexity of the aging process. This implies shifting from disease-centered paradigms towards holistic evaluations of the individual, probably focused on the measurement of objective functions. In this chapter, we provide an overview of current issues affecting research on aging and possible solutions that researchers of tomorrow may valorize.

**Keywords** Intrinsic capacity · Multimorbidity · Alzheimer trials · Aging research future

## 18.1 Research on Aging: An Intricate Matter

The aging phenomenon, defined as the gradual and continuous accumulation of spontaneous changes in the structure and function of materials over time, is a long-lasting matter of scientific investigation. Nevertheless, to date, the mechanisms underlying this process are still largely unknown [1]. Although several gerontologists propose that the key factors regulating the age-related decline of the organisms

M. Cesari (✉)
Geriatric Unit, Fondazione IRCCS Ca' Granda - Ospedale Maggiore Policlinico, University of Milan, Milan, Italy
e-mail: macesari@gmail.com

M. Canevelli
Department of Human Neuroscience, "Sapienza" University, Rome, Italy
e-mail: marco.canevelli@gmail.com

M. U. Pérez-Zepeda
Department of Geriatric Epidemiology, National Institute of Geriatrics, Mexico City, Mexico
e-mail: mperez@inger.gob.mx



C. García-Peña et al. (eds.), *Aging Research - Methodological Issues*,
https://doi.org/10.1007/978-3-319-95387-8_18

will soon be identified to become targets of future anti-aging interventions [2], it is a fact that we do not still exactly know how we age.

However, we have hints on how to limit and manage the consequences of aging. After all, life expectancy has been steadily increasing over time. People today experience a substantially healthier life compared to contemporaries of even few decades ago [3, 4]. Wider access to education, larger availability of healthcare services, and improved medical interventions are surely the cornerstones for such progresses.

It is noteworthy that the aging of the population is not a demographic phenomenon peculiar of high income countries, but a significant modification of societies worldwide, even in the poorest regions of the world [5], whose presence in the scientific literature is extremely scarce. Thus, beyond being complex, research aging is also substantially biased by socio-economic factors pervading the analysis of individuals under study.

## 18.2 Challenges of Research on Aging

The older the person gets, the higher will be the level of his/her biological complexity (and somehow the more challenging will become his/her socio-economic context). The complexity of the older adults is determined by the multi-systemic modifications occurring with aging and clinical conditions (more or less identified). Moreover, positive and negative interactions among diseases, among diseases and medications, and among medications (just to mention few of the many possibilities) further complicate the analysis of the aging organism.

Therefore, one of the major risks in the study of aging is its hyper-simplification around an (often arbitrarily selected and) isolated pathophysiological pathway. It is not uncommon to see researchers seeking for "THE" biomarker able to explain a disease of old age and serve as target for intervention. This approach, typical of a medical model based on stand-alone diseases, cannot obviously work in the complex scenario of older adults.

## 18.3 Alzheimer's Disease Research: A Negative Example

A typical example might be provided by the unsuccessful story of Alzheimer's disease research. To date, the last medication approved by regulatory agencies for this dramatic disease was memantine, released on the market in 2001. In other words, in the last 17 years, nothing new has been proposed as pharmacological intervention, and 99.6% of the conducted interventional studies has reported negative results [6]. The picture is even more worrisome if we consider that we are not talking of a disease-modifier, but simply of an arguable symptomatic medication.

In these last years, many drugs directed against the pathological mechanism identified as responsible for the Alzheimer's disease (i.e., beta-amyloid accumulation) have failed. Why is that? It is possible that arguable methodological choices might have been taken over the years in this field of research. Alzheimer's disease has often been explored as a stand-alone disease, a condition caused by a single and very specific mechanism. Accordingly, it has mostly been intended as sharply separated from the aging process. However, without adequately taking into account that the disease occurs in the context of the age-related accumulation of deficits, the populations of patients enrolled in clinical trials were selected on the basis of stringent eligibility criteria with consequent limitations in the external validity of the findings [7]. The narrow and limited approach to the study of the disease falls in the middle of the long-lasting evidence-based medicine issue in geriatric medicine [8].

At the same time, too much emphasis has perhaps been given to beta-amyloid plaques [9], whose meaning is still very uncertain and predictive value relatively modest [10]. It has not even taken into account the evidence showing that the same beta-amyloid seem to have a quite heterogeneous molecular structure [11], suggesting that the brain amyloid deposition may represent the umbrella under which many biological clusters may exist. It is also worth to be mentioned that some risk conditions (e.g., mild cognitive impairment) have usually been studied as fatally ending into dementia, whereas they may easily and spontaneously reverse to normality [12]. Moreover, there is the worrisome trend to generate novel risk conditions (frequently based on arbitrary assumptions) with the pretext of using them in the preventive research field, and then becoming studying/considering them as "real" diseases (e.g., subjective cognitive decline, cognitive frailty) [13, 14].

These limitations, here exemplified as characterizing the Alzheimer's disease research, can be easily found also in the study of other age-related conditions [15]. They need to be solved, and this can be done only by approaching aging and its consequences with a comprehensive/holistic approach respectful of the complexity of the theme, and without leaving behind the importance of the socio-economic context.

## 18.4 The Study of Aging is in the Complexity

If we assume that aging is a complex phenomenon, the solution for its effects cannot be found in mono-dimensional and "one size fits all" strategies. Under this perspective, there has been an increasing interest around concepts as multimorbidity (i.e., the simultaneous presence of two or more diseases in the individual) and multidimensional interventions. However, these models still do not solve the problem of the age-related complexity and may even seem in contradiction with some of its features [16]. In fact, adding diseases or components of intervention does not guarantee the conduction of a comprehensive and organized approach, but may result into fragmentation of care and working in silos.

In the study of aging, 1 plus 1 never gives 2 as implicitly intended by the sterile multimorbidity construct. The simple addition of the diseases cannot be sufficient for describing the subclinical, complex, and often vague manifestations presented by the older person. The count of detected diseases will never be able to overcome the socioeconomic barriers of the diagnostic process or the arbitrary decisions hidden in the definition of a clinical condition designed for young or adult individuals. Multimorbidity does not clearly take into account the positive and negative interactions potentially existing among diseases [17]. It is flatted over a linear trend in the association with the risk of negative outcomes, whereas age-related deficits tend to accumulate following exponential patterns. Indeed, the field of aging research may be considered as nested in the difference between the linear and exponential increase of risk for negative outcomes depicted by the count of stand-alone diseases and the accumulation of deficits. In that delta, there is a still largely unexplored field of research which should be aimed at 1) disentangling interactions between known diseases, and 2) potentially generating novel conditions from the clustering of signs, symptoms, diseases, and disabilities. Such approach may mean that, for example, hypercholesterolemia is a different condition when associated with diabetes compared to when it is a comorbidity of arthritis. Its meaning and relevance might change because the biological background is different. This shift of paradigm is exactly in line with a system biology approach and personalization of care [16]. In addition, the personal dimension of how a disease or co-existing diseases are perceived by the individual adds a relevant but often neglected piece to the aging research puzzle.

Consistently, multidimensional strategies of interventions might be argued when designed as the simple addition of two or more components. This approach mirroring the idea that "two is better than one" is clearly in contrast with the geriatric "less is more" theory or obvious public health considerations. The multidimensional approach can exist only if coordinated and adapted to the needs, priorities, and resources of the aging organism/individual. The intervention per se does not necessarily need to be multidimensional. It is the initial evaluation that has to be comprehensive and explore the multiple dimensions of the older person's health status in order to provide the correct list of priorities upon which the intervention will be designed. It is noteworthy that this model is at the basis of the comprehensive geriatric assessment, the strongest evidence-based intervention existing in geriatrics. It is perhaps the template we should use for testing novel strategies for age-related conditions even at pre-clinical level or to improve the methodology of our trials. For example, the study of the biology of an age-related disease should not rely on single biomarkers but try to organize the information coming from multiple biological pathways and test the many interactions across molecules, systems, and mechanisms. Or, it might be important to start putting apart the obsolete models of traditional clinical trials in order to privilege those pragmatic designs where interventions are adapted to the biological and clinical features of the individual and tested in the real-life situations of a given healthcare system.

One last consideration in this field should be made on the general idea that screening and anticipating diagnoses is always good. For sure, early detection of diseases is an important asset of preventive medicine and should be promoted. Nevertheless, the conduction of a screening procedure should always adhere to well-established rules [18]. In particular, screening is justified when the benefits coming from the early identification of the abnormality are superior to the consequences of managing the detected condition (both for the individual and the society).

In this context, it is important to underline how the screening of subclinical conditions in frail older individuals may often create more troubles than benefits [19]. Such obvious rule is often and increasingly neglected, perhaps due to the general fear of becoming old. Nevertheless, research on aging cannot forget cornerstones of preventive medicine and consequently act within acceptable boundaries (i.e., increasing physical activity prevents a number of conditions even in the oldest old).

## 18.5 Aging Research at the Individual Level

The model that more than others has the characteristics for promoting the study of aging across species is probably the one proposed by Rockwood and Mitnitski [20]. It is designed to mathematically condense the complexity of aging into a variable, the so-called Frailty Index (FI). The FI is based on the assumption that every living species tends to accumulate deficits as it ages. The quantification of the burden of deficits gives life to an estimate of the biological age of the individual, overcoming the limitations of traditional standards (e.g., chronological age, multimorbidity). The FI, defined by the ratio between the deficits a person presents and the number of deficits explored in the context of a comprehensive geriatric assessment [21], is a strong predictor of negative outcomes [22], can be translated across species (thus promoting translational research) [23], and is suitable for retrospective analyses giving value to databases originally built for different reasons [24–26]. Moreover, the assessment of the risk profile of the individual in terms of percentage of deficits the person presents implicitly allows a standardization of the language in the field of aging. In this way, interdisciplinary and across settings comparisons/discussions can be established favoring the integration of information and valorization of backgrounds/experiences [27, 28].

Although the FI has increasingly been used over the years, many aspects of it are still not adequately explored. For example, the backward process of deconstructing the FI into clusters of deficits for defining subgroups of persons has not yet been frequently applied. Since the FI is generated by a critical mass of information stemming from a comprehensive evaluation of the individual, analyses aimed at seeing how the single items interact in the generation of the frailty status might be extremely interesting. It is a matter of seeking the real foundations of geriatric medicine in the exploration of interactions established by the items generating the biological vulnerability of the person. It potentially means giving life to the definition of novel clinical conditions and therapeutical targets. Overcoming the limitations of traditional nosological conditions, the conduction of these cluster analyses may indicate which deficits are more prone to interact with others and redesign the monodimensional way we see diseases. In other words, it might allow to understand whether the same deficit (e.g., a disease or a symptoms) assumes a different clinical and therapeutical condition according to the interactions with other deficits simultaneously presented by the individual [17]. Has hypercholesterolemia the same meaning when associated with diabetes or Alzheimer's disease? Is it possible that a certain condition might play a different role in the organism's homeostasis according to the corollary of deficits surrounding it?

## 18.6 Aging Research in Public Health

There is an urgent need to reshape the healthcare systems in order to render them more responsive to the needs and priorities of our aging societies. In this context, public health authorities have been paying special attention during the last years to initiatives promoting successful and healthy aging [29–31].

### *18.6.1 Intrinsic Capacity*

In September 2015, the World Health Organization (WHO) released the World Report on Ageing and Health [29]. The document presented a new framework for global action specifically aimed at sustaining the development of new models of healthcare that may be better responsive of the older people's needs. In particular, the report introduced and defined the following key concepts:

- Healthy Ageing: the process of developing and maintaining the functional ability that enables well-being in older age;
- Functional ability: the health-related attributes that enable people to be and to do what they have reason to value. Functional ability is composed by intrinsic capacity, environment, and the interactions between the two;
- Intrinsic capacity: the composite of all the physical and mental capacities of an individual;
- Environment: all the factors in the extrinsic world that form the context of an individual's life.

Recent work has specifically focused on the concept of intrinsic capacity. This construct is designed to capture the biological reserves of the individual, independently of diseases and chronological age. It is the sum of all the key functions (i.e., locomotion, cognition, psychological, vitality, sensory) allowing the individual to properly interact in the environment for meeting his/her objectives [32]. As such, intrinsic capacity follows as a life-long trajectory, declining with aging. Strategies should thus be aimed at keeping the trajectory as closest as possible to the functional ability one and limit the barriers posed by the environment.

This model implicitly takes into account the complexity of diagnosing at advanced age, when the biological substratum is complicated by multiple subclinical and clinical factors [16]. By shifting paradigm from diagnoses to function, intrinsic capacity allows to:

- More pragmatically evaluate the individual, measuring what is probably more important for guaranteeing independent life;

- Assume a life-course approach in the assessment of the person, privileging decisions taken on the trajectory of the health condition rather than on arbitrarily decided defining cut-points;
- To overcome the social barriers hampering the diagnostic process (especially in disadvantaged regions/populations), standardizing the assessment of the individual to his/her biology;
- To give responsibility to the individual about his/her health status;
- To give a positive connotation to the aging process (thus fighting ageism) because talks about functional reserves rather than deficits.

Today, it is perhaps too early to understand whether this model will obtain wide dissemination and adoption. An objective tool for standardizing clinicians and researchers around this language is not yet available. Moreover, it might be perceived a possible competition in the acquisition of the intrinsic capacity model in those settings and specialties where similar comprehensive constructs (e.g., the FI) might already exist. Under this perspective, it is important to understand that the WHO is not aiming at replacing successful initiatives already in place to promote healthy aging. Its aim is to change approach, language and discussions on aging that today tend to exclude older individuals from the society and promote a holistic assessment of the older people's needs. For this reason, the implementation of the model might be easier and faster than what can be imagined, especially in those disciplines (as geriatrics) where the comprehensive assessment of the individual and the personalization of care are well-established cornerstones of the daily practice. For sure, a homogenization of terminology (often used in an inappropriate way) is required to facilitate further advancements in the field.

One additional point is also worth to be clarified. Today, when we talk about conducting a comprehensive assessment of the individual, we automatically think at the filling of multiple scales, tests, and questionnaire on a paper support. Differently, the model of intrinsic capacity is specifically designed for benefiting of novel and future technologies. It is a fact that during our daily life, our functions are continuously registered by numerous devices (e.g., smartwatches, smartphones, or other wearable technologies). The organization of these data can easily feed public health strategies, and support clinicians at individualizing care for their patients. For example, data about physical activity collected by the actimeters included in mobile phones have already been used for describing the geographic distribution of healthy behaviors across countries worldwide [33]. Consistently, when this information is used at the individual level, the amount of physical activity recorded by the device has been used for soliciting the person at doing more and more [34]. Apps like these able to measure other functional domains are also available or under development. Therefore, it will be possible in the future to obtain a comprehensive evaluation of the individual based on the longitudinal tracking of functions without special effort (i.e., continuous registration of potential components of intrinsic capacity). The organization of these data and provision to healthcare professionals may then serve for identifying individuals experiencing a deviation from normality of their intrinsic capacity trajectory and promptly intervene or verify the efficacy of implemented interventions over time.

### 18.6.2 Integrated Care

The WHO published in October 2017 a document on Integrated Care for Older People. These recommendations are strongly related to the World Report on Ageing and Health and aimed at promoting the comprehensive assessment of the individual in order to personalize care through a multidisciplinary approach. This is not really new for the geriatric field. However, it probably represents a legitimation of what geriatricians have been doing for the past decades to optimize care of older adults, and a first attempt to disseminate it to a wide spectrum of health and social care professionals.

The complexity of aging is today affecting the clinical routine of many specialties beyond the geriatric medicine borders. Sometimes, geriatric syndromes are evident in individuals affected by specific conditions causing an acceleration and accentuation of the aging process (e.g., HIV) [35]. In order to find and implement adequate care for the increasing number of frail individuals, it is necessary to develop interactions across disciplines and specialties [36].

Finding solutions for integrating care and promoting multidisciplinary exchanges is one of the next challenges of our health and social care systems. The success of the new models will be surely based on the correct identification of individuals to introduce in the adapted pathways and the cost-effectiveness of the system. In order to guarantee the respect of these pre-requisite, an important contribution might, again, be given by internet and communication technologies. In fact, the centralization of data and merging of clinical and health economic information are crucial for reorganizing the collapsing systems [37].

Recently, in the attempt of providing better care to frail older persons, the Frailty Index has been introduced in the United Kingdom primary care service by contract [38]. The systematic use of this tool has benefited by an already existing centralization of clinical data. As shown in preliminary reports [39], the use of these automatically generated information about the health status of the individual with economic data coming from public health sources might help at optimizing costs and investments. The interaction of the health economics world with the new-to-come medicine based on holistic models is still at beginning, but probably represents the future way to go for making sure that the new paradigms will be sustainable.

## 18.7 Conclusions

In conclusion, aging research is at the edge of massive transformation. Traditional models of care are shifting towards complex multidimensional models. Consequently, research is called at modifying its traditional approach to the theme of aging, abandoning mono-dimensional and disease-specific considerations in favor of holistic evaluations of functions. In this context, it is necessary to promote and develop exchanges and interactions across specialties and disciplines, accepting the inadequacy of independently working methodologies. Only through integration of knowledge and care, it will be possible to address the multiple challenges that our aging society is today facing.

## References


1. Cesari M, Vellas B, Gambassi G (2013) The stress of aging. Exp Gerontol 48(4):451–456. https://doi.org/10.1016/j.exger.2012.10.004
2. de Grey AD (2011) Progress but speed is of the essence. Rejuvenation Res 14(4):351–352. https://doi.org/10.1089/rej.2011.1235
3. Manton KG, Gu X, Lowrimore GR (2008) Cohort changes in active life expectancy in the U.S. elderly population: experience from the 1982–2004 National Long-Term Care Survey. J Gerontol B Psychol Sci Soc Sci 63(5):S269–S281
4. Hayflick L (2000) The future of ageing. Nature 9(408):267–269. https://doi.org/10.1038/35041709
5. Teguo MT, Kuate-Tegueu C, Dartigues JF, Cesari M (2015) Frailty in sub-Saharan Africa. Lancet 385(9983):2151. https://doi.org/10.1016/S0140-6736(15)61021-2
6. Cummings JL, Morstorf T, Zhong K (2014) Alzheimer's disease drug-development pipeline: few candidates, frequent failures. Alzheimers Res Ther 6(4):37. https://doi.org/10.1186/alzrt269
7. Canevelli M, Trebbastoni A, Quarata F, D'Antonio F, Cesari M, de Lena C et al (2017) External validity of randomized controlled trials on Alzheimer's disease: the biases of frailty and biological aging. Front Neurol 8:628. https://doi.org/10.3389/fneur.2017.00628
8. Cherubini A, Del Signore S, Ouslander J, Semla T, Michel JP (2010) Fighting against age discrimination in clinical trials. J Am Geriatr Soc 58(9):1791–1796. https://doi.org/10.1111/j.1532-5415.2010.03032.x
9. Canevelli M, Bruno G, Cesari M (2017) The sterile controversy on the amyloid cascade hypothesis. Neurosci Biobehav Rev 83:472–473. https://doi.org/10.1016/j.neubiorev.2017.09.015
10. Tsartsalis S, Xekardaki A, Hof PR, Kovari E, Bouras C (2017) Early Alzheimer-type lesions in cognitively normal subjects. Neurobiol Aging 62:34–44. https://doi.org/10.1016/j.neurobiolaging.2017.10.002
11. Qiang W, Yau WM, Lu JX, Collinge J, Tycko R (2017) Structural variation in amyloid-β fibrils from Alzheimer's disease clinical subtypes. Nature 541(7636):217–221. https://doi.org/10.1038/nature20814
12. Canevelli M, Grande G, Lacorte E, Quarchioni E, Cesari M, Mariani C et al (2016) Spontaneous reversion of mild cognitive impairment to normal cognition: a systematic review of literature and meta-analysis. J Am Med Dir Assoc 17(10):943–948. https://doi.org/10.1016/j.jamda.2016.06.020
13. Canevelli M, Blasimme A, Vanacore N, Bruno G, Cesari M (2014) Issues about the use of subjective cognitive decline in Alzheimer's disease research. Alzheimers Dement 10(6):881–882. https://doi.org/10.1016/j.jalz.2014.07.154
14. Sánchez-Garrido N, Cesari M, Sgaravatti A, Zengarini E, Moreira V, Borda MG, Zuniga-Gil C et al (2016) The chimeric nihilism of geriatrics. J Am Geriatr Soc 64(11):e213–e214. https://doi.org/10.1111/jgs.14348
15. Canevelli M, Bruno G, Remiddi F, Vico C, Lacorte E, Vanacore N, Cesari M (2017) Spontaneous reversion of clinical conditions measuring the risk profile of the individual: from frailty to mild cognitive impairment. Front Med 4:184. https://doi.org/10.3389/fmed.2017.00184
16. Cesari M, Pérez-Zepeda MU, Marzetti E (2017) Frailty and multimorbidity: different ways of thinking about geriatrics. J Am Med Dir Assoc 18(4):361–364. https://doi.org/10.1016/j.jamda.2016.12.086

17. Gassmann D, Cheetham M, Siebenhuener K, Holzer BM, Meindi-Fridez C, Hildenbrand FF et al (2017) The multimorbidity interaction severity index (MISI): a proof of concept study. Medicine 96(8):e6144. https://doi.org/10.1097/MD.0000000000006144
18. Wilson JMG, Jungner G (1968) Principles and practice of screening for disease. World Health Organization, Geneva
19. Clarfield AM (2010) Screening in frail older people: an ounce of prevention or a pound of trouble? J Am Geriatr Soc 58(10):2016–2021. https://doi.org/10.1111/j.1532-5415.2010.03070.x
20. Mitnitski AB, Mogilner AJ, Rockwood K (2001) Accumulation of deficits as a proxy measure of aging. ScientificWorldJournal 1:323–336. https://doi.org/10.1100/tsw.2001.58
21. Cesari M, Gambassi G, van Kan GA, Vellas B (2014) The frailty phenotype and the frailty index: different instruments for different purposes. Age Ageing 43(1):10–12. https://doi.org/10.1093/ageing/aft160
22. Clegg A, Young J, Illiffe S, Rikkert MO, Rockwood K (2013) Frailty in elderly people. Lancet 381(9868):752–762. https://doi.org/10.1016/S0140-6736(12)62167-9
23. Rockwood K, Blodgett JM, Theou O, Sun MH, Feridooni HA, Mitnitski A, Rose RA et al (2017) A frailty index based on deficit accumulation quantifies mortality risk in humans and in mice. Sci Rep 7:43068. https://doi.org/10.1038/srep43068
24. Hubbard RE, Peel NM, Samantha M, Gray LC, Fries BE, Mitnitski A, Rockwood K (2015) Derivation of a frailty index from the interRAI acute care instrument. BMC Geriatr 15:27. https://doi.org/10.1186/s12877-015-0026-z
25. Pérez-Zepeda MU, Cesari M, García-Peña C (2016) Predictive value of frailty indices for adverse outcomes in older adults. Rev Investig Clin 68(2):92–98
26. Pérez-Zepeda MU, Cesari M, Carrillo-Vega MF, Salinas-Escudero G, Tella-Vega P, García-Peña C (2017) A frailty index from next-of-kin data: a cross-sectional analysis from the Mexican health and aging study. Biomed Res Int 2017:6069374–6069376. https://doi.org/10.1155/2017/6069374
27. Guaraldi G, Brothers TD, Zona S, Stentarelli C, Carli F, Malagoli A, Santoro A et al (2015) A frailty index predicts survival and incident multimorbidity independent of markers of HIV disease severity. AIDS 29(13):1633–1641. https://doi.org/10.1097/QAD.0000000000000753
28. Schoufour JD, Mitnitski A, Rockwood K, Evenhuis HM, Echteld MA (2013) Development of a frailty index for older people with intellectual disabilities: results from the HA-ID study. Res Dev Disabil 34(5):1541–1555. https://doi.org/10.1016/j.ridd.2013.01.029
29. World report on ageing and health (2015) World Health Organization. Switzerland, Geneva
30. European Commission (2011) Synthesis report on the public consultation on the European innovation partnership on active and healthy ageing
31. Bousquet J, Malva J, Nogues M, Manas LR, Vellas B, Farrell J et al (2015) Operational definition of active and healthy aging (AHA): the European innovation partnership (EIP) on AHA reference site questionnaire: Montpellier October 20–21, 2014, Lisbon July 2, 2015. J Am Med Dir Assoc 16(12):1020–1026. https://doi.org/10.1016/j.jamda.2015.09.004
32. Cesari M, Araujo De Carvalho I, Amuthavalli Thiyagarajan J, Cooper C, Martin FC, Reginster JY, Vellas B, Beard JR (2018) Evidence for the domains supporting the construct of intrinsic capacity. J Gerontol A Biol Sci Med Sci. https://doi.org/10.1093/gerona/gly011
33. Althoff T, Sosic R, Hicks JL, King AC, Delp SL, Leskovec J (2017) Large-scale physical activity data reveal worldwide activity inequality. Nature 547(7663):336–339. https://doi.org/10.1038/nature23018
34. Snyder A, Colvin B, Gammack JK (2011) Pedometer use increases daily steps and functional status in older adults. J Am Med Dir Assoc 12(8):590–594. https://doi.org/10.1016/j.jamda.2010.06.007
35. Guaraldi G, Rockwood K (2017) Geriatric-HIV medicine is born. Clin Infect Dis 65(3):507–509. https://doi.org/10.1093/cid/cix316

36. Royal College of General Practitioners & British Geriatrics Society (2017) Integrated care for older people with frailty. Innovative approaches in practice, London
37. Royal College of Physicians (2012) Hospital on the edge? The time for action
38. Clegg A, Bates C, Young J, Ryan R, Nichols L, Ann Teale E, Mohammed MA et al (2016) Development and validation of an electronic frailty index using routine primary care electronic health record data. Age Ageing 45(3):353–360. https://doi.org/10.1093/ageing/afw039
39. Cesari M, Costa N, Hoogendijk EO, Vellas B, Canevelli M, Perez-Zepeda MU (2016) How the frailty index may support the allocation of health care resources: an example from the INCUR study. J Am Med Dir Assoc 17(5):448–450. https://doi.org/10.1016/j.jamda.2016.02.007

# Index

**A**
Abduction, 14, 15, 21
Acetylation, 35, 36
Age groups, 3, 6, 64, 65, 71, 100, 101, 106, 121
Age-related diseases, 56, 58–60
Aging, 2, 18, 26, 55, 64, 76, 78, 91, 96, 123, 136, 160–162, 175, 192–194, 198, 210, 222, 231–239
Aging process, 2, 19, 26, 30, 38, 41, 42, 55, 58, 96, 99, 103, 194, 210, 214, 233, 237, 238
Alzheimer's disease (AD), 57, 222, 223, 225, 232, 235
Animal models, 7, 30–34, 41, 45, 194

**B**
Baseline evaluation, 101
Bias, 4, 15, 75–77, 80, 87–91, 100, 108, 117, 119, 144, 146, 147, 149, 152, 153, 167, 168, 206, 225
Big data, 186–195
Biology of aging, 56, 194
Biomarker, 19, 27, 41, 43, 58–59, 105, 106, 232, 234
Biomedical, 26–45, 59, 157, 188, 190, 192
Biomedical research, 26–45, 58, 189, 191, 192, 225

**C**
Cancer, 5, 6, 18, 34, 38, 39, 55–57, 84, 85, 89, 90, 120, 122, 201, 214, 222
Caregivers, 78, 117, 118, 139, 170, 177, 179, 193
Case-case studies, 86
Case-cohort studies, 85
Case-control, 84–91, 98
Case-control studies, 84–91
Case series, 63, 65, 69–70
Case study, 64, 65, 67–69, 74, 165
Causality, 64, 71, 98, 101, 117
Cause-effect, 19, 84, 100
Cellular, 20, 21, 26, 29, 33, 37, 41–44, 56–58, 225
Cellular senescence, 26, 40–44, 57, 58
Certainty, 15, 16, 21, 148, 182
Chromatin, 34–37
Chronic disease, 2, 56, 58, 59, 66, 90, 121, 158, 161, 166, 194, 203, 214, 222
Clinical observation, 64
Clinical practice guidelines, 146
Clinical trial, 45, 96, 102, 116, 139, 144, 145, 148, 149, 153, 154, 163, 200, 211, 212, 214, 233, 234
Cognitive decline, 103, 107, 233
Cognitive deterioration, 101, 103
Cognitive function, 40, 102, 103
Cohort design, 98, 103


C. García-Peña et al. (eds.), *Aging Research - Methodological Issues*,
https://doi.org/10.1007/978-3-319-95387-8

Common sense, 12, 15, 20
Complexity, 18, 19, 35, 44, 99, 160, 169, 183, 193, 194, 207, 232–236, 238
Confounding factor, 15, 153
Costs, 90, 102, 103, 124, 146, 169, 170, 198, 200, 201, 205, 224, 238
Credibility, 15, 78, 80, 137, 206
Cross-sectional studies, 65, 163, 164

**D**

Databases, 64, 147, 164, 187, 190, 195, 235
Data collection, 13, 74–79, 91, 101, 119, 132, 139, 148, 159, 163, 165, 186, 187, 190
Data mining, 186–195
Decision-makers, 159, 163, 171, 198, 205, 206
Deductive logic/deduction, 12, 14, 15, 21, 130
Deficiency, 32, 33
Dementia, 2, 4–6, 38, 57, 64, 103, 122, 136, 170, 177, 179, 222, 233
Demographic data, 148, 166
Dependence, 5, 6, 103, 124, 211, 218
Deprescription, 122
Depression, 98, 130, 134, 136–138, 164, 168
Descriptive, 63–71, 130, 136, 150, 163, 164, 166, 172
Diagnostic test studies, 89
Diaries, 75, 77
Disability, 1, 2, 57, 59, 100, 103, 106, 107, 158, 178, 200, 218, 222
DNA-damage, 29–34
DNA-damage repair syndromes, 30
DNA methylation, 34, 35, 37
DNA repair, 29, 30, 32–34
Documents, 75, 77, 90, 144, 145, 165, 194, 199

**E**

Ecological studies, 64, 66, 70
Ecological validity, 178, 183
Economic evaluation, 163, 169, 170
Effectiveness, 6, 102, 117, 119, 124, 145, 150, 153, 159, 165, 166, 169, 202, 203, 205, 207
Efficiency, 146, 159, 203, 204
Electronic, 147, 164, 190, 192
Emergency, 66, 68–70, 166
Emergent diseases, 21, 78
Emergentism, 17, 21
Epidemiological studies, 57, 70, 91, 96, 175, 183, 190
Epigenetics of aging alteration, 34–35, 103
Ethnography, 74
Evidence-based, 117, 125, 145, 146, 166, 198, 222, 233
Experimentation, 12, 124
External validity, 65, 118, 139, 165, 233

**F**

Falls, 6, 66, 70, 71, 99, 103, 124, 178, 233
Focal groups, 79
Frailty, 2, 4–7, 18, 19, 21, 39, 57, 64, 66, 96, 101, 103, 107, 120, 122–124, 176–178, 200, 212, 216, 218, 233, 235, 238
Frailty syndrome, 39, 178
Full disclosure, 15, 16
Function, 2–6, 26–32, 34, 40, 44, 57, 59, 68, 121, 123, 160, 211, 224, 231, 236–238
Funding, 124, 222–228
Funnel plots, 152

**G**

Genes, 32, 36, 37, 39, 44, 191
Geographic, 107, 177, 179, 237
Geriatric population, 107
Geriatric syndromes, 2, 4, 64, 66, 101, 118, 124, 200, 238
Gerontechnology, 116
Geroscience, 55–60, 123, 194
Government agendas, 199
Grants, 223–225, 228
Grounded theory, 74, 78

**H**

Hallmarks of aging, 26, 38, 45, 56, 58
Health-care policies, 144
Health research, 3, 139, 158, 189, 193, 217
Health systems research (HSR), 64, 78, 157–172
Healthspan, 38, 40, 56
Heterochronic parabiosis, 28, 40
Heterochronic transplantation, 29
Histone, 34–37
History of scientific method, 12
Human aging, 17–20, 30, 35, 110
Human capital, 202
Human premature aging disorders, 31–32
Hypothesis, 14–16, 18–21, 130, 164, 193

## I


Impact evaluation (IE), 159, 166–168
Implementation research, 119, 159, 163, 165
Incident cases, 87
Independence, 6, 167, 207, 216
In-depth interviews, 75, 76, 79, 166
Induction, 12, 14, 15, 21
Inflammaging, 39
Inflammation, 26, 39–41, 55–57, 69, 103
Information, 1, 12, 65, 74, 86, 96, 117, 132, 144, 162, 176, 186, 200, 227, 234
Interventions, 2, 39, 56, 57, 64, 90, 102, 110, 116–118, 121–124, 131, 133, 137–139, 144, 146, 148, 158, 159, 162, 163, 165, 166, 169, 170, 175, 177, 183, 189, 192, 198, 200, 201, 203–207, 222, 232–234, 237
Intrinsic capacity, 236, 237


## K


Kidney cancer, 89
Knowledge, 2–4, 8, 12, 13, 16–18, 21, 26, 57, 59, 64, 66, 74, 77, 80, 91, 96, 101, 102, 120, 137, 144, 157–159, 162, 171, 182, 186–189, 192–194, 198, 205–207, 215, 224, 239


## L


Lifespan, 26, 33–36, 38, 39, 41, 44, 56, 194, 226
Logical empiricism, 116
Logical reasoning, 15
Longevity, 19, 43, 45, 96, 97, 101, 104, 209
Longitudinal methodology, 103, 110
Longitudinal observation, 100
Longitudinal studies, 7, 65, 96–110, 164, 182, 226


## M


Mammals, 35, 36, 39, 40, 44, 45
Mechanistic target of rapamycin (mTOR), 38, 39
Meta-analysis, 144–154, 163, 201
Methodology, 1, 98, 101, 103, 110, 131, 148, 234
Methods, 3, 12, 40, 73, 84, 108, 130, 144, 158, 176, 188, 206
Mixed methods, 74, 130–139, 166
Mobile app, 237
Mobile phones, 176–180, 182, 186, 193, 237
Model of aging, 37, 226
Monitoring studies, 67, 70, 71
Morbidity-mortality, 103, 107, 146
Multimorbidity, 116, 118, 233, 235
Muscle, 26, 28, 29, 40, 57–59, 86, 123, 178
Muscle strength, 57, 58, 67


## N


Narrative phenomenology, 74
Narrative reviews, 65, 145
National Institute on Aging (NIA), 56, 102, 110, 223
National Institutes of Health (NIH), 189, 223, 224
Neurodegenerative diseases, 57, 58, 222
Non-pharmacological interventions, 116, 117, 121, 123
Novel, 26, 40, 41, 43–44, 64, 158, 159, 177, 183, 186–195, 222, 233–235, 237
Nutrient sensing, 26, 37–39, 58


## O


Observation, 12–15, 20, 41, 63, 64, 75–77, 79–80, 88, 89, 98, 100, 109, 130, 138, 139, 163, 177, 212
Observational studies, 79, 100, 144, 169, 212
Odds ratio (OR), 86, 150, 151
Older adults, 1, 19, 39, 57, 64, 74, 84, 96, 116, 134, 144, 158, 178, 193, 199, 211, 222, 232
Osteoarthritis, 69
Osteoporosis, 31–34, 57, 59, 84
Outcomes, 2, 4, 6–8, 19, 20, 38, 64, 70, 99, 100, 103, 104, 118, 119, 123–124, 130, 132, 134, 137–139, 146, 148, 150, 158, 159, 166, 167, 169, 178, 183, 190, 191, 201, 227, 234, 235


## P


Panel studies, 97, 103
Paradigm, 2, 13, 160, 186, 193, 234, 236, 238
Parkinson's disease (PD), 57, 58, 177, 225
Pharmaceutical industry, 118, 122, 190, 215
Pharmacological clinical trials, 116, 120, 122
Philosophers, 13, 14, 20, 213
Philosophy of science, 12–13, 17
Physical activity, 57, 98, 116, 118, 123, 176–178, 203, 235, 237
Physical health, 103
Physiological aging (PA), 30–32
Political-normative, 198
Political viability, 202, 205

Polypharmacy, 102, 118, 121
Positivism, 130
Pragmatists, 15
Prediction, 12–15, 20, 97, 187–189
Premature aging, 30–34
Presuppositions (P), 15, 16
Prevalence, 18, 19, 64–67, 90, 91, 97, 118, 130, 136, 158, 161, 192, 222
Prevalence studies, 64–66
Prevalent cases, 87
Progeroid syndromes, 31, 32, 34, 37
Prospective, 84, 90, 98, 117, 120, 166, 228
Prostate cancer, 84
Proteostasis, 26, 56–58
Public agenda, 199
Public policies, 3, 144, 160, 165, 166, 197–207
Publication bias, 152, 153

**Q**
Qualitative, 12, 43, 73–81, 130–139, 149–151, 163, 165, 166, 205
Qualitative research, 73–81, 133, 139, 163
Qualitative studies, 78, 80, 136, 138
Quantitative methods, 80, 130–132, 135, 137–139, 166

**R**
Rapamycin, 38
Reductionism, 17, 21, 130
Relative risk (RR), 100, 150
Reliability, 91, 144, 151, 154
Representativity, 74
Research, 2, 12, 26, 56, 63, 73, 84, 96, 116, 130, 144, 157, 175, 188, 198, 209, 222, 232
Retirement, 97, 104, 105, 168, 224
Risk difference, 150

**S**
Sample unit, 108, 110
Sampling strategies, 65, 74, 75
Sarcopenia, 32, 57–59, 66–68, 86, 123, 124, 200
Scientific knowledge, 12, 13, 16, 158, 215
Scientific method, 3, 12, 144, 193
Scientific progress, 13
Scientific revolution, 12
Semi-structured interviews, 76, 136, 137, 139, 165
Senescence, 26, 27, 35, 37, 40–44, 57, 58
Senescence associated secretory phenotype (SASP), 39–44
Shelterin complex, 42
Sirtuins, 35, 36
Smartphone, 177–179
Statistical data, 200
Stem cells (SCs), 26–29, 34, 40, 41, 56, 57, 59, 123, 124, 215
Structured interviews, 76
Surveillance studies, 64, 66, 70–71
Systematic review, 102, 144–154, 163, 170
Systemic lupus erythematosus, 64, 68, 89

**T**
Taxonomy, 14, 15, 20, 66, 71
Technology, 1, 3, 116, 123, 154, 175–183, 189–191, 193, 207, 222, 224, 227
Telomere, 26, 29–34, 41, 42, 58
Telomere attrition, 26, 29–34, 42, 58
Therapeutic interventions, 39, 116, 144, 146, 148
Therapeutic targets, 35, 41
Transparency, 12, 15, 16
Trend designs, 97

**U**
Ubiquitous sensing technology, 175–183

**V**
Variable-dependent, 14
Variable-independent, 14

MIX
Papier aus verantwortungsvollen Quellen
Paper from responsible sources
FSC® C105338

If you have any concerns about our products,
you can contact us on
**ProductSafety@springernature.com**

In case Publisher is established outside the EU,
the EU authorized representative is:
**Springer Nature Customer Service Center GmbH**
**Europaplatz 3, 69115 Heidelberg, Germany**

Printed by Libri Plureos GmbH
in Hamburg, Germany